John O'Reilly

The Placenta

the organic nervous system, the blood, the oxygen, and the animal nervous system,

physiologically examined

John O'Reilly

The Placenta
the organic nervous system, the blood, the oxygen, and the animal nervous system, physiologically examined

ISBN/EAN: 9783337392857

Printed in Europe, USA, Canada, Australia, Japan

Cover: Foto ©berggeist007 / pixelio.de

More available books at **www.hansebooks.com**

THE PLACENTA,

THE

ORGANIC NERVOUS SYSTEM,

THE BLOOD,

THE OXYGEN,

AND

THE ANIMAL NERVOUS SYSTEM,

PHYSIOLOGICALLY EXAMINED.

BY

JOHN O'REILLY, M.D.,

LICENTIATE AND FELLOW OF THE ROYAL COLLEGE OF SURGEONS IN IRELAND ;
RESIDENT FELLOW OF THE NEW YORK ACADEMY OF MEDICINE ; MEMBER
OF THE MEDICO-CHIRURGICAL COLLEGE OF NEW YORK ; LATE
MEDICAL OFFICER TO THE OLDCASTLE WORKHOUSE AND
FEVER HOSPITAL, IRELAND.

" Quidquid præcipies, esto brevis, ut cito dicta
Percipiant animi dociles, teneantque fideles."—Hor.

NEW YORK:

S. S. & W. WOOD, 389 BROADWAY.

LONDON:

JOHN CHURCHILL, NEW BURLINGTON STREET.

1861.

FIRST EDITION.

TO

ROBERT ADAMS, M.D., A.M., M.R.I.A.,

PRESIDENT OF THE ROYAL COLLEGE OF SURGEONS IN IRELAND,

THIS VOLUME

IS INSCRIBED,

AS A TOKEN OF RESPECT AND ESTEEM,

BY

THE AUTHOR.

TO

VALENTINE MOTT, M.D., LL.D.,

EX-PRESIDENT OF THE MEDICAL FACULTY OF THE UNIVERSITY OF NEW YORK, EMERITUS
PROFESSOR OF SURGERY IN THE UNIVERSITY OF NEW YORK, ETC., ETC.

MY DEAR SIR:

In the estimation of the public, and by the universal consent of
the Profession, you stand unrivaled as a Surgeon in the United States.
Your great achievements in operative surgery will perpetuate your
name as long as surgery is cultivated as a science.

In dedicating the second edition of this volume to you, I do not ex-
pect to elevate your character; I am simply actuated by the motive
of recording my deep and everlasting gratitude for the great solicitude,
kindness, attention, and skill evinced by you towards me while laboring under two attacks of illness which imperiled my life in the first
years of my residence in this city.

I have the honor to be,
With the highest respect and esteem,
Yours most faithfully,

JOHN O'REILLY.

230 WASHINGTON SQUARE, SOUTH,
NEW YORK, 21ts. *November*, 1861

INTRODUCTION.

In issuing a second edition of this work, I have to remark, the arrangement is altogether changed : deficiencies are filled up, errors are corrected, *new* and *interesting* matter is added—the theories advanced heretofore are shown to be founded on facts, as well as on the authorities of men whose testimony cannot be doubted.

It is to be hoped that, notwithstanding the subjects treated of are very abstruse and very difficult of comprehension, yet that they are made clear and intelligible.

PREFACE.

THE importance of being thoroughly acquainted with the laws which regulate the functions of the *Organic Nervous System*, the *Blood*, the *Oxygen*, and the *Animal Nervous System*, cannot be too forcibly inculcated. The practice of Medicine and Surgery, to command success and win the admiration of mankind, must be placed on a thoroughly scientific foundation. It is an imperative duty on every member of the Profession, whether in a high or low position, to uphold the honor and dignity of his profession, and contribute, as far as in him lies, to the advancement of the various branches of knowledge connected with it.

In treating of the subjects which come under discussion in this volume, I candidly confess my incompetency to do them the justice their vast importance demands. I hope and believe, however, I have put forward in a sufficiently prominent manner my ideas about them in such a way as to attract the attention of persons possessed of greater talents and higher scientific attainments. I have to plead as an apology for the style, the arrangement, and the imperfections of the work, that I was engaged as a general practitioner in Medicine

and Surgery for nearly eighteen years before my attention was particularly directed to the subjects now considered ; and that during that period, as well as up to
the present time, I have been engaged in an extensive
and laborious practice, affording but little time for
study or the cultivation of scientific acquirements.

Having been fully impressed, by the teachings of the
esteemed Professors whose Lectures I had the advantage of attending whilst a student, that a medical man
should be thoroughly educated in every department
of the Profession, I thought it would be derogatory to
the character of the School I represented if I listened
in silence to opinions I could not *endorse*, or approved
of a theory by an affirmative act I believed to be
wrong. To adopt either of these alternatives would
be demonstrative of a *lack* of *moral courage*.

The manner in which I vindicated my views under
such circumstances will be understood on an attentive
perusal of all the subjects contained in this work,
which I confess cost me some thinking, and a little
study. My labors, however, will be fully compensated
for, in the event of the Profession deriving advantage
from the study of the matters brought under consideration.

CONTENTS.

THE PLACENTA.

ORGANIC NERVOUS SYSTEM.

10

ANIMAL NERVOUS SYSTEM.

13

14

VARIOUS KINDS OF BATHS.

MODUS PROPAGANDI OF THE HUMAN SPECIES.

SYPHILITIC POISONING OF THE ORGANIC NERVOUS SYSTEM.

THE PLACENTA.

In order to demonstrate that the organic nerves surrounding the maternal uterine arteries inosculate with the organic nerves surrounding the hypogastric arteries of the fœtus, thus establishing a nervous communication between mother and child, it becomes necessary to study the anatomy of the placenta, by its analogy to other organs, as well as the comparative anatomy of the Invertebrata in relation to the organic nervous system.

1. That the placenta resembles a conglomerate gland.

2. That the placenta bears a very strong resemblance to the *liver* in its anatomical organization and its function.

3. That the placenta is composed of four sets of vessels, connected together by cellular tissue.

4. That two sets of vessels enter the placenta in large trunks—namely, the maternal uterine arteries, which convey the arterial blood from the mother to the placenta; and the hypogastric arteries, which carry back the blood to the placenta, after its circulation in the fœtus.

5. That two sets of vessels commence in capillaries—namely, the umbilical vein, which conveys the arterial blood to the fœtus; and the uterine veins, which commence in capillaries, and proceed directly to enter the decidua uteri, and discharge their contents into the uterine sinuses, and through the latter to the venous circulation of the mother.

2

6. That the uterine arteries subdivide into innumerable branches, ultimately terminating in capillaries; and that the hypogastric arteries subdivide into innumerable branches, and terminate in capillaries.

7. That retinæ of organic nerves encompass and form plexuses round the arteries, sending twigs into their coats, (viz., the uterine and hypogastric arteries.)

8. That the organic nerves of the maternal uterine arteries form glands at the capillary terminations of the arteries; and that the organic nerves of the hypogastric arteries also form glands.

9. That the glands formed at the capillaries of the hypogastric and uterine arteries inosculate, and freely communicate.

10. That the blood is arterialized by the combined influence of the organic nerves derived from the mother and fœtus in the placental lobule or gland.

11. That the blood which has been arterialized is conveyed from the glands or lobules by the capillaries to the branches and trunk of the umbilical vein to the fœtus; and that the impure blood is carried back by the capillary maternal uterine veins from the glands to the uterine sinuses.

12. That the four sets of vessels in the liver consist of the hepatic arteries, which are analogous to the maternal uterine arteries; the vena porta,* which is analogous to the hypogastric artery of the fœtus; the hepatic veins, which are analogous to the branches of the umbilical veins; the gall-ducts, which are analogous to the uterine veins.

13. That the analogy is still further made more manifest by knowing that the hypogastric arteries contain the impure blood which has circulated through the fœtus, and that the vena porta contains the impure blood which has circulated through the intestines; that the umbilical vein contains the blood which has been purified for the fœtus, and that the hepatic veins contain the blood after having been purified from the bile; that the uterine veins carry back all the impurities of the fœtal blood to the uterine sinuses, and are analogous to the gall-

---

* The vena porta is *surrounded* by a *retina of organic nerves;* a fasciculus of the plexus which surrounds the hepatic artery is distributed to the vena porta.

ducts, which convey the impurities of the intestinal blood to the gall-bladder.

14. That the vena porta subdivides into innumerable branches, and terminates in capillaries; that the hepatic veins commence in capillaries; that the hepatic artery terminates in capillaries: that the gall-ducts commence in capillaries.

15. That the four sets of vessels in the liver are all connected at their capillaries, terminating in the acini or lobules, in the same manner as the four sets of vessels are in the placental lobule.

16. That it is admitted that the liver plays a prominent part in the fœtus as regards the purification of the blood for the fœtus, and in this function resembles the placenta.

17. That the manner in which the uterine veins enter the decidua uteri, to pour their contents into the uterine sinuses, is analogous to the mode in which the cerebral veins enter the dura mater, to discharge their contents into the cerebral sinuses.

18. That the uterine sinuses are situated between the walls of the uterus and the decidua uteri.

19. That the *bruit de soufflet* may be adduced as a proof that the uterine veins pierce the decidua, to pour their contents into the uterine sinuses, inasmuch as this phenomenon is produced by the blood flowing from constricted orifices into expansive openings.

20. That Professor Simpson's operation of removing the placenta, in placental presentation, affords a proof that the uterine veins pierce the decidua to reach the uterine sinuses, inasmuch as its removal cuts off the supply of venous blood from the sinuses.

21. That the blood discharged in placental presentation is of a venous character, showing that it is the product of the uterine veins, which have pierced the decidua to reach the uterine sinuses.

22. That when the placenta is situated over the os uteri, as soon as the latter is dilated to the size of a shilling, there is profuse hæmorrhage. Here the placenta is separated from the walls of the uterus. The placenta can be felt with the finger, and here it is manifest the uterine veins must pierce the decidua to reach the uterine sinuses; the blood, under the circumstances

stated, flows from the mouths of the veins directly through the expanded os, and escapes per vaginam; the uterine sinuses being ruptured during the dilatation of the os uteri.

23. That Professor Dalton says it is very easy to demonstrate the arrangement of the *fœtal tufts* in the human placenta. They can be readily seen by the naked eye, and may be easily traced from their attachment at the *under surface* of the chorion to their termination near the *uterine* surface of the placenta. The fœtal tufts are *the maternal uterine veins* proceeding from the inner to the external or uterine surface of the placenta, to pierce the decidua, and discharge their contents into the uterine sinuses.

24. That the chorion, which surrounds the fœtal vessels, on their entrance into the placenta, is analogous to the capsule of Glisson, which surrounds the hepatic vessels on their entrance into the liver.

25. That the distribution of the maternal uterine artery, on passing through the placenta to reach its internal surface, is analogous to the renal artery, which passes to the cortical substance of the kidney.

26. That mental impressions made on a pregnant woman are communicated to the fœtus in utero, inasmuch as the animal and organic nervous systems of the mother act in concert, whereby the impression is communicated by the mother to the fœtus in utero, through the inosculation of the organic nerves of mother and fœtus in the placental lobule.

27. That when it happens that a woman has witnessed a frightful accident befall her husband during the last months of pregnancy, she will give birth either to a dead child, or one who in due time will prove to be an idiot, or suffer from some nervous affection.

28. That Professor Owens' description of the fœtus and appendages of the kangaroo demonstrates that the fœtal vessels terminate as well as commence in capillaries; thus confirming my views relative to the anatomical character of the placenta.

29. That there is every reason to believe that the oxygen of the maternal arterial blood is united with the venous blood of the fœtus by the action of electricity.

A STATEMENT OF DR. JOHN BURK,

OF GRAND STREET, NEW YORK,

ON THE COLOR OF THE BLOOD IN HÆMORRHAGE FROM THE UTERUS BEFORE OR AFTER DELIVERY.

ACCORDING to my observation, and I have met with a great number of cases, the blood is of the venous character, like that which is removed by venesection, though frequently much darker.

I remember well the case of a Mrs. Grey, who, being seven months pregnant, was seized immediately after dinner with intense pain in the right side. When I saw her the pain was very severe; the uterus could be felt through the abdominal walls, as hard as a stone. By-the-by, according to my observation, a peculiar stony-hardness of the womb, attended with intense local pain on either side, and great prostration, are sure indications of internal hæmorrhage. No dilatation of the os tincæ. She died in 24 hours after the first symptoms of pain.

Post-Mortem.—Placenta on the right side. It was detached about two inches from the uterine wall, superiorly; about three pints of dark fluid blood were effused between the uterine wall and membrane.

I have observed in other cases of uterine hæmorrhage dark masses of coagulated blood to come away immediately after the expulsion of the child, and sometimes before it.

[Professor Barker, of this city, told me the blood in placental presentation was dark-colored, or venous. The observations of Dr. Burk and Professor Barker confirm my views, that the uterine veins pierce the decidua to reach the uterine sinuses, and that the blood must be dark-colored, or venous.]

ORGANIC NERVOUS SYSTEM.

LIFE

" Is a power or imponderable agent,"* located or centred in the organic nervous system, capable of preserving the body from decomposition as long as it continues in the organic nervous system, but requiring oxygen in order to make itself manifest, or rather the operations which characterize life manifest, viz., respiration, circulation, and animal heat.

Life is imparted to the semen at the instant of its emission, as more fully explained elsewhere.

Life may continue dormant for a shorter or longer period; many persons have remained for days in a state of ecstasy or trance, apparently lifeless, yet have again resumed the functions of life.†

---

* Coldstream on Animal Electricity.

† PERSONS APPARENTLY DEAD.

Pliny, " De his qui elati revixerunt," is quoted by Dr. T. M. Beatty in the first volume of the " Cyclopædia of Practical Medicine," page 548, who, amongst other instances, gives that of the Roman Consul, Avicula, who, being supposed dead, was conveyed to his funeral-pile, where he was reanimated by the flames, and loudly called for succor, but before he could be saved was enveloped by the fire, and suffocated. Bruhier, a French physician, who wrote on the uncertainty of the signs of death, (1742,) relates an instance of a young woman upon whose supposed corpse an anatomical examination was about to be made, when the first stroke of the scalpel revealed the truth; she recovered, and lived many years afterwards. The case related by Philippe Pue is somewhat similar: he proceeded to perform the Cæsarian Section upon a woman who had, to all appearance, died, undelivered, when the first incision betrayed the awful fallacy under which he acted. A remarkable instance of resuscitation after apparent

A person after being immersed in water for some minutes, and to all appearance dead, will again have respiration, circulation, and animal heat established.

A frog may be frozen and thus resemble an icicle, yet, on being thawed, will leap about as friskily as before.

It is manifest from what has been just stated, that life is an independent power or agent, and that it can hold possession of its abode, located in the organic nervous system, for a certain time, independently, or without the presence of either the oxygen or the blood; a matter fully and conclusively established by the fact that life will remain in the organic nervous system for some time after all the blood is removed, and consequently, the oxygen with it.

Life, or instinct, is the invisible power under whose guidance

---

death occurred in France, in the neighborhood of Douai, in the year 1745. and is related by Rigeaudeaux, (*Journal des Sçavans*,) in 1749, to whom the case was confided. He was summoned in the morning to attend a woman in labor at a distance of about a league: on his arrival, he was informed that she had died in a convulsive fit two hours previously. The body was already prepared for interment. On examination, he could discover no indications of life. The os uteri was sufficiently dilated to enable him to turn the child and deliver by the feet. The child appeared dead also, but by persevering in the means of resuscitation three hours, they excited some signs of vitality, which encouraged them to proceed, and their endeavors were ultimately crowned with complete success. Rigeaudeaux again carefully examined the mother, and was confirmed in the belief of her death, but he found that, although she had been in that state for seven hours, her limbs retained their flexibility. Stimulants were applied in vain, and he took his leave, recommending that the interment should be deferred until the flexibility was lost; at five P. M. a messenger came to inform him that she had revived at half past three. The mother and child were both alive three years after.

There is scarcely a dissecting-room that has not some traditional story handed down, of subjects restored to life after being deposited within its walls. Many of these are mere inventions to catch the ever greedy ear of curiosity; but some of them are, we fear, too well founded to admit of much doubt. To this class belong the circumstances related by Louis, the celebrated French writer on Medical Jurisprudence. A patient who was supposed to have died in the Hospital Salpétrière was removed to his dissecting-room; next morning Louis was informed that moans had been heard in the theatre, and on proceeding thither, he found, to his terror, that the supposed corpse had revived during the night, and had actually died in the struggles to disengage herself from the winding-sheet in which she was enveloped; this was evident from the distorted attitude in which the body was found.

the complicated and ingenious organization of man carries on its various functions, removing from the body what is deleterious to it, and retaining what is useful to it; protecting organs when threatened with danger, and repairing organs or parts of the body which have suffered injury. Life, located in the organic nervous system, merely requires oxygen and blood to perform these wonderful phenomena and exhibit its supernatural wisdom; or, in other words, discharges processes which cannot be imitated by human ingenuity.

Description of the Organic Nervous System.

Bichat says, " The ganglions have a color very different from that of the nerves. They present a soft, spongy tissue, somewhat similar to the lymphatic glands, but which has nothing in common either with the cerebral substance or that of the nerves."

Mr. Quain states: "Each ganglion is invested by a thin investment of cellular tissue which connects it with the surrounding structures, and forms a nidus for the vessels which supply its proper substance; but when this is removed, *another lamella* is found forming an envelope for the filaments which enter and depart from the *ganglion*, as *well* as for that body itself."

Mr. Harrison says, "This capsule is surrounded by areolar tissue and blood-vessels; the latter ramify on and pierce the capsule; the internal surface of the latter is very vascular, and may on the *larger ganglions* be separated as a *vascular membrane* from the external fibrous layer, and is *analogous* to the *pia mater* on the *cerebro-spinal* axis. The *mass* of a *ganglion* is composed of a *plexus* of *nervous filaments*, with a variable quantity of vesicular or gray neurine; the afferent nerves divide into numerous fibrillæ, which pass in the most *varied directions* and reunite most probably in different combinations, the interstices being filled with capillary vessels and gray neurine."

Scarpa applies the term "*nervi molles*" to the tissue of the organic nervous tissue, in consequence of the *softness of its structure*. The analogy between the descriptive anatomy of the brain and organic ganglion is worthy of attention: the

former is invested with a fibrous membrane, called the dura
mater; the latter is also invested with a fibrous membrane,
serving for the same purpose to the ganglion as the dura mater
does to the brain; the brain is invested with a vascular membrane, called the pia mater; the ganglion has an investing
membrane of a similar kind; the brain is composed of nervous
tubules, with white and gray substance; the ganglion is composed of nervous filaments, with a variable quantity of vesicular or gray neurine. The cerebral glands furnish a volatile
agent to enable the brain to carry on the operations of animal
life, whilst the arteries distributed to the vascular membrane
of the ganglion afford oxygen to the nervous filaments of the
ganglion, enabling it to discharge its vital functions.

Bichat considered each ganglion a distinct nervous centre,
which proved him to be a man of great sagacity and penetra-
tion.

Mr. R. D. Grainger states, " The ganglions of the great sympathetic, consisting of the following, connected on each side of
the body with what is regarded as the trunk of the nerve, viz.,
three cervical, twelve dorsal, three to four or five lumbar, three
to five sacral; to these we must add some large masses placed
near the mesian line plane, viz., two semilunar, three or four
cœliac ganglions, and one cardiac ganglion, first described by
Wrisberg, but which is occasionally deficient; and lastly, forming a part of the great sympathetic, the ophthalmic, the spheno-
palatine, the otic, and the submaxillary ganglions, and a
small body usually met with in the cavernous—the cavernous
ganglion. M. Hip. Cloquet has described in rather vague
terms a small reddish mass placed in the anterior palatine
canal, which he calls the *naso-palatine* ganglion."

As Mr. Grainger says the cardiac ganglion " is occasionally
deficient," I shall quote other authorities to prove its importance, and consequently, its existence invariably.

" The cardiac ganglion," says Mr. Quain, " lies beneath the
arch of the aorta; and the bifurcation of the trachea in close
contact with the former, extending from the division of the
pulmonary artery to the origin of the brachio-cephalic. This
may be considered as the common point of union of the cardiac
nerves that issue from the cervical ganglions, and the immedi-

ate source from which the different nerves proceed to supply
the heart."

Scarpa thus describes the cardiac ganglion: "*Anastamosis
illa, valde insignis, quæ inter utriusque lateris cardiacorum ner-
vorum truncos, sub aorta curvatura, paulo supra cor conficitur.*"

Mr. Harrison remarks: "The size and structure of the cardiac ganglion are very variable. *Instead* of *a single distinct*
ganglion, it often appears as *a congeries* of *small ganglia* en-
tangled in the plexus of the uniting nerves."

I have called the congeries of ganglions described by Mr.
Harrison according to the functions of the organs with which
they are connected, viz.: the right auricular ganglion, the right
ventricular ganglion, the left auricular ganglion, the left ventricular ganglion, the aortic and pulmonary ganglions.

Mr. Harrison says, in describing the semilunar ganglions,
" Instead of a single mass, they consist of a congeries of knot-
ted ganglions on the nervous cord." I have named these ganglions according to the functions they have to perform, viz.:
the phrenic ganglion, the gastric ganglion, the hepatic ganglion, the renal ganglion, the lacteal or mesenteric ganglion,
the pancreatic ganglion.

The tenth and eleventh dorsal ganglions, from which the
lesser splanchnic nerve arises, by two roots, and form one
trunk on the last dorsal vertebra, and then join the renal
plexus, should be called the *spermatic ganglions;* inasmuch as
the spermatic artery receives its plexus of nerves from the renal
plexus; and inasmuch as the propagation of the species is a
distinct and specific function, requiring distinct vital ganglions
to preside over its highly important functions.

I have added two other ganglions, not hitherto, I believe,
described as such: namely, the Pineal gland, which I have named
the superior central ganglion; and the pituitary gland, which I
have named the cerebral ganglion.

Pineal Gland, or Superior Central Ganglion.

The superior central ganglion is in direct communication
with the brain, through its peduncles, or two bands of white
matter. It is necessary that the ganglion should be so at-

tached, inasmuch as it regulates the action of the cerebral gland, located in the gray substance of the brain. As the white substance of the brain is in connection with the gray at every part, it follows as a sequence, that the ganglion being connected with the white substance of the brain in the manner stated, it must be in connection with the gray substance. The ganglion regulates the functions of the brain, inducing *sleep* when the brain *requires* repose, by suspending the action of the cerebral glands, and consequently, arresting the evolution of the volatile agent given off to the nerve-tubules of the brain. The ganglion communicates impressions to the cerebral glands located in the gray substance, through the nerve-fibres or tubules; to the otic ganglion, through the facial nerve; to the lenticular ganglion, through the communicating branch of the third nerve; to the cardiac ganglion, through the par vagum; to the pulmonary ganglion, through the par vagum; to the semilunar, the phrenic, gastric, hepatic and renal, through the par vagum; to the spermatic, through the spinal cord and roots of the eleventh and twelfth dorsal nerves; to the cervical dorsal lumbar and sacral nerves, through the spinal cord and roots of spinal nerves; to the spheno-palatine, through a branch of the second division of the fifth nerve. The cerebral ganglion will be more particularly described in another place.*

TO RECAPITULATE

The Functions of the Organic Ganglions and Glands.

The superior central ganglion regulates the functions of the cerebral glands. and thus *protects* the brain from *overstimulation* or *overwork*, by *arresting* the action of the *cerebral glands*, and thus inducing sleep.

The *cerebral* ganglion *regulates* the nutrition, assimilation, absorption, and secretion of the brain.

The *carotid ganglion regulates* the force of the circulation

---

* I have called the pituitary gland the *cerebral ganglion*. It presides over the functions of nutrition, assimilation, secretion, and absorption in the brain.

28

through the arteries of the brain, and thus *prevents* destruction of the organ by *too* great an impetus of the blood.*

The lenticular ganglion guards, protects, and regulates the function of the eye, so as to meet the requirements of the mind, by regulating the contraction and relaxation of the iris.

The otic ganglion regulates the action of the organ of hearing, by regulating the contraction and relaxation of the *tensor tympani* muscle.

The spheno-palatine ganglion presides over the functions of the induction of food into the stomach, including salivation, mastication, deglutition, as well as the function for drinking fluids to allay thirst, when there is too much oxygen in the blood.

The superior cervical ganglion presides over the function of the intonation of the voice, as well as certain muscles to which it gives branches.

It is also concerned in the operations or actions of the heart.

The middle regulates the functions of the thyroid glands, the muscles to which it sends nerves, and is concerned in the movements of the heart.

The inferior cervical ganglion regulates the actions of the muscles to which it sends branches, as well as regulates the secretion of milk in the female.

The cardiac and pulmonary ganglions, taken collectively, regulate the functions of the heart and lungs, as more particularly explained elsewhere.

The semilunar, the hepatic, the diaphragmatic, the sphenic, the gastric, the renal, the mesenteric, preside over the secretion of the gastric juice, the secretion of the bile, the contraction and relaxation of the diaphragm, the secretion of the urine, the contraction of the intestinal tube, and the process of absorption by the lacteals as well as lymphatics.

The spermatic preside over the secretion of the semen.

The vertebral ganglions preside over the contraction and relaxation of the muscles.

All the ganglions hold communication with one another, and the organic nerves derived from the ganglions distributed

---

* A fact proved by vivisection by myself, in presence of others.

over the body are connected with the animal nerves; a complete interlacement and inosculation of the organic and animal nerves take place all over the surface of the body.

The cerebral glands secrete the volatile or phosphoric agent.

The pulmonary glands secrete electricity, or vital fluid.

The gastric glands secrete the gastric juice.

The hepatic glands secrete the bile.

The renal glands secrete the urine.

The spermatic glands secrete the semen.

The cutaneous glands secrete the sweat, or cutaneous perspiration.

The salivary glands secrete the saliva.

The mammary glands secrete the milk.

Other glands in different parts discharge the functions al-
lotted to them; as, for instance, the glands in the intestinal tube
secrete the serum or mucus, under the operation of certain
agents.

Nervous System of the Invertebrata.

In some of the lowest classes of animals, very great difficulty
has been experienced by anatomists in discovering a nervous
system of any kind.

Trebly, Goede, and Carus failed to discover any trace of a
nervous system in the Acalepha.

Dr. Grant describes a nervous system which he found in the
Boroe Pileus, consisting of a double circular nervous filament,
situated round the oval extremity of the body, which sends off
minute filaments in each of the spaces between the longitudinal
bands of ciliæ.

These eight points, from which the longitudinal filaments
come off, present ganglionic enlargements.

Spix, a German anatomist, describes a nervous system in the
Actinia, which may be considered an isolated polypus, having
no calcareous skeleton, and fixing itself in the rocks by its
fleshy base, consisting of minute filaments, with minute ganglia
surrounding the fleshy base, from which were given off nerves
to the different parts.

Amongst the Echinodermata, Tiedeman describes in a small

species of this genus a nervous system, consisting of a circular cord around the mouth, from which proceeded a filament along each ray. having at its origin a minute ganglionic enlargement. The nervous ring rested upon the extreme edge of the central aperture in the calcareous frame-work of the body; and the filaments rested on the inferior surface of the rays, concealed by, and at the base of, tubular feet and suckers.

Two other filaments, much shorter than those just described, are given off from each of these ganglionic enlargements, to be distributed to the stomach and other viscera.

This animal possesses considerable muscular power.

In the Ascidia Mammillata, belonging to the Mollusca Tunicata. Cuvier describes and figures the nervous system as consisting of a single oblong ganglion, situated near the anus of the animal, and between *that* and the bronchial orifice. From the ganglion branches are given off; some of which, passing to the œsophagus, encompass it in the form of a ring.

This animal is surrounded by a muscular sac, which, by its contraction, can compress and empty its general cavity. This receives some muscular filaments.

The solitary ganglion of the Ascidia seems to regulate the action of the orifices of ingestion and egestion, and of its enveloping sac, on which depends the slight locomotive action of the free species.

Dr. Anderson says, in the Conchifera the nervous system is adapted for the functions these animals have to perform, which are: ingestion of the food, respiration, and locomotion.

These nervous centres, or ganglia, are consequently placed in immediate relation to the organs destined to those functions.

The œsophageal and labial ganglions are the most important. They are two in number, situated more or less near the mouth, and are united by a transverse band, which arches over it.

From these ganglia nerves are given off to the mouth and tentacles, and to the anterior parts of the viscera.

Each ganglion has a branch of communication to the pedal ganglion and to the bronchial ganglion.

In the Distoma Hepaticum, belonging to the Entozoa, Bogannus describes a nervous system, consisting of a nervous collar or ring, with two lateral ganglia entwining the œsophagus;

and two nerves, which are distributed to the posterior part of the body.

Otto describes the nervous system of the Strongylus Gigas as consisting of median nervous filaments, with closely approximated ganglia.

The Cirrhopoda have abdominal cords, with ganglia developed on them; and there is a nervous collar around the œsophagus.

The Annelida have a varied number of ganglia, united by double longitudinal fissures.

In the Crustacea, the common Talitrus has a regular series of ganglia developed at an equal distance from each other.

In the Myriapoda, the Scolopendra Morsitans has a nervous system consisting of twenty-one double ganglia, situated on the ventral surface of the body, connected by intervening double longitudinal cords. From each ganglion are given off lateral nerves, to supply the neighboring muscles, viscera, and feet. These ganglia are nearly all of an equal size, excepting the first, which is the largest, and from which are given off additional nerves to supply the maxilla.

Mr. Owens says each joint of the Articulata corresponds to a division of the nervous system.

In the Gastropoda, the common snail has two nervous centres: one placed *above* the œsophagus, the other below it—both connected by two cords, embracing the œsophageal tube. The upper ganglion supplies nerves to the muscles of the mouth, as well as the skin in its vicinity.

It likewise furnishes the nerves of touch, and of vision, besides those distributed to the generative organs. And from the sub-œsophageal ganglion, which fully equals the brain in size, arise those nerves which supply the muscles of the body and viscera.

The nervous centres obey the movements of the mass of the mouth, with which they are intimately connected. They are pulled backward and forward by the muscles, serving for the *protrusion* and *retraction* of the oral apparatus, and are thus constantly changing their relations with the surrounding parts.

In the Snail, it would seem that the great mass of the nervous collar which embraces the œsophagus will, in some instances,

32

permit the mass of the mouth to pass entirely through it; so that sometimes the brain rests on the œsophagus, and at other times it is placed on the inverted lips.

In the Nudibranchiate, the nervous centres exist in the most concentrated form; and, indeed, it is doubtful whether there are any other ganglia, excepting the large supra-œsophageal ganglion.

In the Tritonia, there are four tubules placed across the commencement of the œsophagus, the nervous collar being completed by a simple cord.

All the nerves which supply the skin, the muscular integument, the tentacles, the eye, and the muscles of the mouth, arise from the tubules; and anatomists have not hitherto detected any other source of supply. For these particulars I am indebted to Dr. T. Rymer Jones.*

---

* Physiology substantiates the truth of the first chapter of Genesis, verse 20: " Let the waters bring forth the creeping creature having life." It is to be remarked, that the creeping creatures were made on the fifth day, and that the nervous system of the Invertebrata was formed for the purpose of giving an animal organization for the manifestations and operations of vitality in these creatures. Geology confirms the truth of the creation; the shells of the creeping creatures are found in the lowest strata of the earth, whilst the fossil remains of the beasts of the earth are found in the upper strata of the earth. The creeping things were made one day, and the cattle and beasts were made on another day; and judging from geology, several thousands of years elapsed between the two days. Again, it is to be remarked, that there is a decided improvement made in the nervous system of the animals that were created on the sixth day; whilst man, who stands at the highest point of the creation, was, according to Genesis, the last animal made. as well as the highest made in point of development of the animal nervous system; on the operations of the latter man, as a rational being, depends. Physiology shows that in the lowest classes of animals, the Invertebrata, there is only an organic or vital nervous system, and that the nervous system becomes more fully developed in the various grades of creation up to man. Geology shows that the creeping things were made before the cattle or the beasts of the earth, inasmuch as the shells of the mollusks are found in the lower strata of the earth; the fossil remains of the megatherium are found in the upper strata of the earth; whilst the skeleton of man is only found on the top of the earth. As some persons may suppose that I am denying the truth of the Book of Genesis, which states the world was made in six days, I beg to state that the Book of Genesis was written by Moses. and that it is not necessary to believe that Moses intended saying that God wrought the creation in six consecutive days, otherwise it would be difficult to understand the 17th verse, 4th chapter, of Genesis: "And Cain knew his wife, and she conceived and brought forth Henoch, *and he built*

To Prove the Identity of the Organic Nervous System of the Invertebrata and the Organic Nervous System of Man.

The next matter should be to determine what division of the nervous system in the Vertebrata is identical with, or discharges the same functions as, the nervous system of the Invertebrata.

All animals require for the continuance of life the ingestion of food, or nutriment, into the stomach. Therefore, in all animals there must be provision made for the function of deglutition, and such is found to exist in all animals.

---

a city, and called the name thereof by the name of his son, Henoch." It is evident that when Moses said he built a city, he meant himself and his offspring. and his offspring's offspring, built a city, which they could easily do during his life, when men lived to be upward of 900 years old. A man gets married at the age of 50 years, and is father of a son or daughter each year up to the age of 450; he lives fifty years after the birth of his last child, or till he is 500 years old. His sons and daughters marry at the ages of 50 and 51 years, and each couple has a son or daughter each year up to the time of the father's death. viz. 500 years. Each couple of the third generation marry at the ages of 50 and 51 years, and have children up to the time of the father's death. Each generation. down to the last, get married at the ages above mentioned, and have children up to the father's death. That being the case, what is the number of the progenitor's offspring at the time of his death? Seven thousand, five hundred and ninety millions, six hundred and forty-three thousand. six hundred and fifty-one.

Supposing that ninety-nine out of each hundred died, there would yet remain seventy-six millions, six hundred and seventy-three thousand, one hundred and sixty-eight.

It would appear an absurdity to state that one man built a city, or that Moses intended saying so; so, in like manner, it is not necessary to say that the creation was made in six consecutive days, or that Moses intended conveying that idea. His saying the evening and the morning was the first day, simply meant that all things specified in the text were made in the time that elapsed between the evening and the morning of each day. On reference to the 5th chapter, verse 31, of Genesis, it will be admitted that persons 450 years old were not too old to propagate the species. "And all the days of Lamech came to seven hundred and seventy-seven years, and he died. And Noe, when he was five hundred years old, begat Sem, Cham, and Japheth." Without further entering into a discussion of what many may suppose is foreign to the present investigation, I cannot help stating that geology, physiology, as well as the construction a person is entitled to put on the first chapter of Genesis after reading the other chapters of Genesis, written by Moses, fully establish the truth of the record of the creation, and confirm the order in which the things were created. It is well to establish the truths of the Bible, when opportunities offer.

3

The nervous rings surrounding the mouths of the lowest classes of animals, such as the boroc pileus, the star-fish, the œsophageal ganglions found in the mollusca, preside over the function of deglutition in the Invertebrata.

The location, the position, the distribution of the nerves, point out the spheno-palatine ganglion as presiding over the function of deglutition in the Vertebrata.

It will be recollected that this is one of the organic ganglia. I will now quote Mr. HARRISON's description of the spheno-palatine ganglion. " It is a small, triangular, reddish substance. It is imbedded in fat, surrounded by branches of the internal maxillary artery, and is situated on the external side of the nasal plate of the palate-bone, which separates it from the cavity of the nose, behind the tuberosity of the superior maxillary bone, and in front of the pterygoid processes. Three sets of branches pass from the ganglion: an inferior, internal, and a posterior.

"First, the inferior, or palatine nerves, descend in the bony canal of that name; send through the canal some small twigs to the spongy bones, and near the palatine separate into three filaments; an anterior, middle, and posterior. The anterior is the largest, and passes forward in a groove within the alveoli, and above the mucous membrane, supplying the latter, the bone and teeth, and finally enters the foramen incisivum by a very fine filament, which communicates with the nerves in the septum narum.

"The middle and posterior filaments of the palatine nerves are distributed to the amygdalæ, the soft palate, and uvula. The posterior usually descends through the osseous canal of the pterygoid portion of the palate-bone. The internal branch of the spheno-palatine nerve is very short, passes through the spheno-palatine hole to the upper and back part of the nose, and divides into five or six branches. The most important of these pass immediately into the mucous membrane covering the superior and middle spongy bones; one branch, called the naso-palatine nerve of Cotunius, passes beneath the sphenoidal sinus, across the root of the nose, and descends obliquely forward, along the septum nasi, as far as the foramen incisivum, where it communicates with the interior palatine branches, and where some anatomists describe a small ganglion (naso-palatine) to exist. This, however, in the human subject, can seldom be distinguished from the surrounding fat and vessels.

"The third, or posterior branch of the ganglion, is the vidian, or superior petrosal nerve: this passes backward through the vidian canal, above the pterygoid plate, and sends some small filaments into the sphenoidal sinuses; it there perforates the cartilaginous substance that closes the foramen lacerum anterius, enters the cranium, and divides into branches—an inferior and posterior. The inferior, or carotid branch, enters the cavernous sinus, and joins the plexus formed around the artery, by the ascending branches of the superior cervical ganglion of the sympathetic. The superior, or petrosal branch, runs backward and outward, beneath the dura mater and Casserian ganglion, in a groove on the petrous bone, enters the hiatus Fallopii in the bone, and becomes attached to the portio dura nerve—the part of function being marked by a small gangliform expansion. The vidian nerve accompanies the portio dura as far as the back part of the tympanum; it then leaves it, receives the name of chorda tympani, and enters the tympanum a little below the pyramid; invested by mucous membrane, it now proceeds forward, between the long leg of the incus and the handle of the malleus—to the latter it is firmly connected; it then escapes by a canal, which appears near the internal extremity of the glenoid fissure; it next runs downward, inward, and forward, joins the gustatory nerve, and continues attached to it as far as the submaxillary gland; it now leaves the gustatory nerve, and unites with filaments from it in the submaxillary ganglion, which is situated near the posterior edge of the submaxillary gland, and from which a number of filaments proceed; these form a plexus, which supplies the gland."

Proofs from Comparative Anatomy that the Œsophageal Ganglion of the Invertebrata is identical with the Spheno-Palatine Ganglion of the Vertebrata.

In the Ruminantia, the spheno-palatine ganglia are very large: they are double the size in a sheep, when compared with a carnivorous animal of similar dimensions. I presume it is unnecessary to specify (the mode of mastication in these animals is quite different from the carnivorous) or to discuss the reasons why the spheno-palatine should be so large in the former, as reflection will at once suggest the answer.

Again, it will be recollected that a snail, having a supra-œsophageal ganglion, has the power of regurgitation.

In fact, it can swallow its oral apparatus, and again regurgitate it. In this respect, it resembles the Ruminantia. It is a remarkable fact that several of the Gastropoda, to which the snail belongs, have stomachs similar to the Ruminantia.

Pathological Proofs

That the Spheno-Palatine Ganglion presides over the Function of selecting proper Food, Mastication, Salivary Secretion, Deglutition, as well as gives the Sensation of Thirst, and is identical with the Œsophageal Ganglion of the Invertebrata.

Hydrophobia.

In one, two, three, or four months—sometimes a longer period elapses—a person who has been bitten by a rabid animal will observe some change in the appearance of the part where the wound was inflicted; either pain, redness, or discoloration of the skin will attract his attention. After some time, the well-known symptoms which characterize hydrophobia will present themselves. The great thirst, with the apprehension of swallowing fluids—the frightful spasm of the glottis, with all the muscles of deglutition, on attempting to swallow—the presence of viscid saliva, which harasses the patient, and which he frequently attempts to spit out—the snapping with the teeth—the rolling of the eyeballs—the paroxysm of strangulation, brought on by hearing water poured from one vessel into another—the paroxysm brought on by looking into a mirror—sometimes the bad odor that annoys the patient, as well as the sudden death of the patient—all demonstrate that the spheno-palatine ganglion is in a morbid state of irritation. The connection and distribution of the nerves of the spheno-palatine ganglion prove conclusively the truth of this affirmation.

Explanation of the Symptoms.

The spheno-palatine ganglion is connected with the superior division of the fifth nerve, (the superior maxillary nerve.) It is connected with the superior cervical ganglion. The spheno-palatine ganglion is connected with the ear by the vidian nerve, and with the submaxillary gland by the chorda tympani nerve. The superior cervical ganglion is connected with the lenticular ganglion by a branch of communication. The spheno-palatine ganglion sends nerves to the muscles concerned in deglutition; sends nerves to the arytenoid muscle; hence the spasm of the muscles can be accounted for. The rolling of the eyes can be accounted for by recollecting the inferior oblique muscle is supplied with a nerve from the lenticular ganglion; the alternate relaxation and contraction of these muscles (inferior oblique) cause the rolling of the eyes. The connection of the vidian nerve with the ear accounts for the bad consequences resulting from certain sounds. The secretion of the viscid saliva is accounted for by the connection of the chorda tympani with the submaxillary gland; the snapping with the teeth is accounted for by the connection of the dental nerves with the spheno-palatine ganglion, through the connection of the superior maxillary nerve, from which the dental nerves take their origin. The sudden dissolution of the patient is caused by spasm of the glottis; the non-admission of air into the lungs accounts for the cause of death—namely, the want of oxygen to combine with the organic nervous glands.

As a further proof that the spheno-palatine ganglion is the organ morbidly affected when hydrophobia is present, it is almost unnecessary to remark, that the saliva in a rabid animal is secreted under the influence of the spheno-palatine ganglion, through the operation of the chorda tympani division of the vidian nerve, which takes its origin from the posterior part of the spheno-palatine ganglion.

Venomous Reptiles.

When a venomous reptile wishes to inflict a deadly wound, it communicates its wishes to the spheno-palatine ganglion, through the superior maxillary division of the fifth nerve: and the latter to the submaxillary glands, through the chorda tympani nerve. (The connection between the animal and organic nervous systems is here made manifest.)

Functions common to Vertebrata and Invertebrata.

Respiration, circulation, and digestion are functions common to the class of animals known as the Invertebrata and the other class called the Vertebrata. The nerves *accompany* the *arteries* to all the viscera and muscles in the Invertebrata. This observation is true of the Vertebrata; the nerves accompany the arteries to the heart, to the lungs, to the liver, to the kidneys, to the testes; distinct nerves are distributed to the iris, to the heart, the intestinal tube, and the uterus.

The Theory that Circulation, Respiration, and Digestion depend on the Operations of the Par Vagi, overthrown.

The vagi may be divided without stopping respiration, circulation, or digestion. LONGET operated on dogs, some of which lived to the fifth day. DUPUY found that horses lived up to the seventh day. DE BLAINVILLE, that pigeons lived to the seventh day.

Irritation of the Vagi does not produce Muscular Excitement of the Heart.

LONGET mentions that he failed in influencing the rhythm of the heart by the application of galvanism to the vagi of dogs, rabbits, and sheep.

Irritation of Cardiac Nerves Influences Rhythm of the Heart.

LONGET very frequently succeeded in influencing the rhythm of the heart by scraping the cervical cardiac branches of the vagus.

Explanation of Phenomena.

It is to be remarked that LONGET did not draw any *distinction* between the *animal* and *organic nerves;* he forgot, or did not appreciate the fact, that the cardiac nerves were derived from the cervical ganglions, and consequently of precisely the *same character* as the nerves *distributed* to the heart, whilst the vagus is an *animal nerve;* hence the different results of the experiments can be explained.

LEGALLOIS has proved by numerous experiments that an animal will continue to breathe after the division of both vagi in the neck, if care be taken to secure the ingress and egress of air to and from the lungs.

Mr. REID observes, that if the vagi be injured above the origin of the recurrent laryngeals, none of the muscles attached to the arytenoid cartilages can any longer act in unison with the muscles of respiration—all these movements cease, and the superior aperture of the larynx can no longer be dilated during inspiration.

Explanation.

The organic nerves which supply the muscles of the larynx are derived from the superior cervical ganglion; the pulmonary plexus is partly formed by filaments from the long cardiac nerve, derived from the superior cervical ganglion; division of the vagi, as above stated, destroys the unity of action between the organic nerves in the larynx and pulmonary plexus.

Life not destroyed by the Removal of the Cerebrum, Cerebellum, and Division of the Vagi.

Mr. REID says, "Although respirations were much diminished by the removal of the cerebrum and cerebellum, and then dividing the vagi, they continued for a longer or shorter time."

VOLKMAN, FLOURENS, and LONGET confirm these observations by experiments.

No Change in the Lung caused by the Tying of the Vagus.

Mr. REID confirms the experiments of Dupuytren, that no morbid change could be discovered in the lungs of dogs, on the side on which the vagus had been tied, in six months after the operation.

Irritation of the Trunks of the Vagi does not cause Muscular Contraction of the Stomach.

Messrs. MAYO and MULLER failed in exciting muscular contraction in the stomach by irritating the trunks of the vagi.

Galvanism.

BICHAT, TIEDEMAN, GMELLEN, LONGET, BRESCHAT, MILNE EDWARDS, inferred that muscular movements can be excited in the stomach of a living animal by galvanizing the lower end of the vagi in the neck, from its effects upon the digestive process.

Fallacy of Experiments.

These gentlemen seem to have forgotten that the branches of the par vagi *inosculate* with *branches* of the *stomachic plexus* in the stomach, and that the *secretion* of the *gastric juice* depends on the operation of the *latter nerves.*

Muscular Movements of Stomach continue after
Section of Vagi.

MAJENDIE observed that these *muscular movements* of the stomach continued after the *section* of the vagi. Mr. REID confirmed MAJENDIE'S remark by experiments on a dog; where, after *cutting* the vagi, and on the dog recovering, he found that the stomach could still *propel* the chyme onward towards the duodenum.

Sensations of Hunger experienced after Division of Vagi.

Messrs. REID and LONGET found that dogs, whose vagi had been divided, *experienced sensations of hunger*, if they survived a certain number of days.

Process of Digestion after Division of Vagi.

LEURET and LASSAIGNE detail the result of an experiment on a horse, where the *process* of *digestion* went on after the *division of the vagi*, with loss of substance.

Experiments on Dogs.

ARNEMAN tied the vagi of a dog, and as the animal *lived until the 165th day* after the operation, it was killed.

SEDILLOT, CHAUMENT, and Mr. REID arrived at similar results from experiments on dogs—that *the digestive process was carried on after the division of the vagi.*

It appears almost unnecessary to remark that the dog, *which lived until the* 165th *day, had all the vital functions duly carried on.*

Action of Organic Nerves.

The contraction and dilatation of the iris; the contraction and dilatation of the heart; the contraction and dilatation of the arteries; the contraction and dilatation of the stomach; the contraction and dilatation of the intestinal tube; the contraction and relaxation of the diaphragm; the contraction and relaxation of the uterus—are attributable to the action of the organic nerves.

The Iris.

The iris is a circular muscle, and receives its nerves exclusively from the lenticular ganglion; its contraction and dilatation is caused by the action of the ciliary nerves, derived from the lenticular ganglion. To give an example of the fact, it is only necessary to observe the pupil of a person on passing from a dark room to one well illuminated, and again returning to the former, when alternate contraction and dilatation of the pupil will be perceived. In the one case it contracts, to preserve the retina from too strong a glare of light; in the other it expands, to allow a greater number of rays of light to enter through the pupil, in order to have objects clearly impinged on the retina.

The Heart.

The several cavities of the heart receive nerves from the cardiac ganglions; alternate contraction and relaxation of the cavities is the result. The same class of nerves is distributed to the heart and the iris.

Arteries.

The arteries are surrounded by plexuses of nerves,* derived from the organic ganglions; twigs or branches of the nerves pierce the coats of the arteries. Alternate contraction and dilatation takes place. The arteries and the iris receive nerves from the same organic nervous system.

* Ample proofs that such is the case will be given in another place.

Intestinal Tube.

The intestinal tube is a circular muscle; it receives nerves from the mesenteric ganglion; it contracts and dilates. The nerves that supply it and the iris belong to the same system.

Uterus.

The uterus is a hollow muscle; it contracts and dilates. It, too, receives its nerves from the same source as the iris.

Having reason to believe that many persons cannot comprehend the *modus operandi* of the organic nerves, and suppose it to be impossible that the nerves should be possessed of such extraordinary power, I will endeavor to explain the difficulty of comprehending the matter, by directing attention to another phenomenon, somewhat analogous in its operations, viz.:

Let two bars of iron be magnetized, and placed parallel to one another. The positive poles, it will be perceived, will repel each other, increasing the distance between the extremities of the bars, (or dilatation.) Again, the north pole of one magnet will be observed to attract the south pole of the other. (or contraction.)

The Mode of Operation of the Brain and Cerebro-Spinal Nerves adduced, to show that there is nothing more extraordinary in the Operations of the Organic Nervous System than is momentarily witnessed in the Operations of the Animal Nervous System.

A person, on being told to move rapidly his fingers and toes, will instantly do so; or he can move one or more, according to circumstances. The impression made on the auditory nerve is extended to the nerves distributed to the fingers and toes, by the act of the will, or volition. The nerves being capable of receiving impressions at their extremities, as well as conveying impressions to their extremities, it follows that the operations of the mind must be *coextensive* with the extremities of the nerves or nerve-tubules. In order to understand this matter, see what happens when a person is told to move his little finger and great-toe at the same time.

An explanation of these phenomena, which would consist in stating that the impression made on the auditory nerve in the

internal ear was transmitted to the brain, and from the brain
dispatched by the spinal cord to the fifth, sixth, seventh, eighth
cervical and first dorsal nerves, to the interlacement of these
nerves in the brachial plexus, to the ulnar nerve and the superficial branch of the ulnar nerve distributed to the little finger;
and further, at the same instant that the dispatch was sent
along the whole course of the spinal cord to the sacral plexus,
to the great sciatic nerve, to the internal division of the sciatic nerve, and the internal plantar nerve distributed to the
great-toe, could be met with numerous objections. As it is
impossible to conceive how the message could be sent, without
confusion to the branches of the spinal nerves, forming the
brachial plexus, or in what manner it could select the ulnar
nerve, that only forms one of the divisions of the brachial
plexus, and finally, that division of the ulnar nerve distributed
to the little finger. Again, how the message could divide
itself, and go down the spinal cord to the branches of the
sacral plexus, the great sciatic nerve, the internal division of
the latter, and lastly, the internal plantar nerve. That there
is a connection between the extremities of the nerves and the
brain, is a fact that admits of easy demonstration; division of
the ulnar or internal plantar arrests the communication. It
appears that the vesicular nervous matter contained in the
interior of the nerve-fibre of the nerve, and the vesicular nervous matter contained in the interior of the nerve-fibre or
tubule of the brain, form a complete chain of communication
between the vesicular matter contained in the nerve-fibres of
the brain and the vesicular matter contained in the nerve-
fibres. That on the operations of the brain and spinal cord
being brought into action by the volatile agent *generated* by
the *cerebral glands, located* in the *gray substance* of the *brain,*
and *similar nervous glands, located* in the *gray substance* of the
spinal cord, the mind can *extend* its *operations* to the *most distant part of the body,* inasmuch as the vesicular nervous matter
contained in the *filaments* of the *nerves distributed* to the *finger* is of *the same character,* and is connected with similar
nervous matter in the nerve-tubules of the brain. The operations of the mind must necessarily *extend* to the tip of the
finger, and the tip of the finger must also be conscious of the

operations of the mind, and capable of communicating with it, as evidenced by the sense of *touch*.*

Other Proofs that the Pineal Gland is a Ganglion of the Organic Nervous System; that it belongs to the same Class of Ganglions as the Lenticular and Superior Cervical Ganglions, &c.

Experiments performed by POURFOUR DU PETIT.

Effects of the Section of the Cervical Sympathetic Nerve.

1. "CONSTRICTION OF THE PUPIL."

2. "THE EYE SEEMS TO BE SMALLER, OR EVEN TRULY SHRINKS."

Explanation of the Phenomena

Caused by POURFOUR DU PETIT'S *Experiments—Section of the Cervical Ganglion causes Irritation of the Superior Cervical Ganglion.*

1. "Constriction of the pupil."

The irritation is propagated to the lenticular ganglion by a branch sent to it by the superior cervical ganglion.

The lenticular sends the ciliary nerves to the iris; the irritation is propagated to the iris, and causes it to contract, precisely as irritation of the Pineal gland, on being irritated, causes contraction of the pupil, through its communication with the lenticular ganglion, by a branch of the third nerve.

2. "The eye seems to be smaller, or even truly shrinks."

The irritation caused by section of the cervical nerve extends to the lenticular ganglion, and from thence to the third or motores oculorum, through communicating branches of the third nerve with the lenticular ganglion; the irritation is propagated to the muscle supplied by the third nerve; spasm of the muscle is the consequence. The ball of the eye must necessarily be drawn backward, and presents the appearance described by POURFOUR DU PETIT. Irritation of the Pineal gland causes rolling of the eyeballs, contraction and dilatation of the pupils,

---

* This matter will be again reverted to.

followed by tetanic spasm of the muscles supplied by the third
nerve, and gives the eyeball a fixed position, similar, I presume, to the one described by POURFOUR DU PETIT.

Proofs adduced from Pathology, to demon-strate that the Pineal Gland is a Ganglion of the Organic Nervous System.

A person may labor under chronic hydrocephalus for years,
and his mental as well as vital faculties continue unimpaired.
This appears very extraordinary, but can be accounted for.
The convolutions are unfolded; the cranium becomes enlarged;
the brain becomes expanded; the ventricles filled with serum;
the Pineal gland or ganglion floats in the fluid. Here, it will
be remembered, the body of the ganglion lies on the tubercula
quadrigemina, and is not attached to them; consequently, as
the ganglion does not suffer from pressure, the functions of
life are not interfered with. A child is attacked with menin-
gitis; all the symptoms of acute inflammation are present; the
eyes glisten and the pupils are contracted, showing that the
central ganglion is suffering from irritation. After some time,
the child is attacked with convulsions, followed by dilated
pupils, partial or complete coma, a quick, feeble, and intermittent pulse. In due time the patient dies; serum is found in
the ventricles. What has taken place? In the commencement, the ganglion was suffering from irritation; towards the
close, from oppression. Previous to the attack, the cavity of
the skull was sufficiently large to accommodate its contents;
but in the latter stage it was too small to contain its contents,
and therefore, the skull being incapable of expansion, the
serum effused pressed on the ganglion.

The suffering of the ganglion from the pressure is communicated to all the other ganglions; hence vitality in the organic
nervous system is gradually impaired, until ultimately it is totally extinguished.

A man may have the side of his head, with a large slice of
the brain, removed by a sabre, and yet apparently sustain no
vital injury, and have his intellectual faculties unimpaired.
Another is pitched on top of his head, and, after some time,

will be found in a state of insensibility, with dilated pupils, stertorous respiration, slow, laborious pulse.

In the former case, although a formidable injury is inflicted, yet it is confined to the animal nervous system; and as one side remains intact, the intellect is not disturbed. No important organ of organic life being interfered with, accounts for vitality continuing undisturbed.

In the latter case, the central ganglion suffers from the pressure of a clot of blood or depressed bone. The suffering of the ganglion from pressure is communicated to the lenticular ganglion, to the cardiac ganglion, and pulmonary ganglion. impairing and diminishing vitality in each of these ganglions; hence the dilated pupils, the slow and laborious action of the heart, and oppressed respiration.

If the pressure is kept up, vitality continues to grow weaker and weaker, until it ceases to exist. It will be remembered, the ciliary nerves are derived from the lenticular ganglion; that the heart receives its nerves from the cardiac ganglion; that the lungs are supplied with nerves from the pulmonary ganglion. It will be further recollected, that the Pineal gland is attached to the inner side of each optic thalamus by a band of white matter; that the brain is in communication with the lenticular ganglion through a branch of the third nerve; that the brain is also in communication with the cardiac and pulmonary ganglions through the connection of the *par vagi*, which freely inosculates with the branches of nerves derived from the cardiac and pulmonary ganglions.

Proofs that Life is located in the Organic Nervous System.

A blow directed to the pit of the stomach will destroy life, by the violence inflicted on the semilunar ganglion, and which is instantly communicated to all the other ganglions, destroying life in all. (A case of this kind is mentioned by SIR A. COOPER, in his Lectures.) A blow on the cardiac ganglion will destroy life in the same way.

A blow on the superior cervical ganglion will cause either death or suspended animation.

A blow in the centre of the forehead will cause death or suspended animation. in consequence of the shock being communicated to the *central superior ganglion, (the Pineal gland.)*

Example.

When a butcher is about killing an ox, he aims the blow at
the central point in the forehead, round where the hair is a
little curled; a smart blow either kills or causes the animal to
fall, when the extremities will be observed to be thrown into
an *extended position*, from *tetanic spasm of the muscles;* the
irritation caused by the violence of the blow is propagated
from the *central ganglion* to all the *other organic ganglions.*

The central spot, above described, is placed opposite the
central ganglion.

An eel, which is so tenacious of life, can be immediately
killed by giving it a smart blow on the tail; the shock, communicated to the organic ganglions in the tail, is communicated
to the other ganglions, and death follows.

Another Proof that Life is in the Organic Nervous System.

An animal, such as a sheep, may have all the blood drawn
off, and yet give evident and vigorous manifestations of life for
some minutes; *showing* conclusively that life is *not located in
the blood.** It is to be further observed, that, as the blood is
being drained off, convulsions set in. and continue at intervals
until death takes place. The convulsions are indications of a
struggle between life and death. The organic nervous system
suffers *irritation*, from the *supply* of the *oxygen* contained in
the blood being *cut off;* and as life cannot exist in the organic
nervous system without the oxygen but for a short time, it puts
forth its whole strength to obtain it before its departure from
its abode, as is witnessed in the violent struggle that takes
place.

---

* John Hunter believed in the vitality of the blood. Moses says, " The *blood*
is the *life* of the flesh." Moses was an inspired physiologist, and his explanation
is true. The blood contains the oxygen, without which life cannot exist in the
organic nervous system.

*Additional Proof that Life is centred in the Organic
Nervous System.*

Every joint of the class of animals known as the Articulata
possesses a distinct nervous system, capable of carrying on **all**
the functions appertaining to its individual capacity. Hence
it is that a lobster may be *partitioned* into several parts, and
each part be *still living.*

The common earth-worm may be divided into two parts,
and *each will be capable of forming a perfect animal.*

A snail may have its head *cut off* and get a *new* one.

Here, as BICHAT truly states, every ganglion is an independent nervous system in itself.

Having already alluded to the tenacity of life in an eel, and
its susceptibility of having its life destroyed by a smart blow
on the tail, in consequence of the shock being communicated to
all of the organic ganglions from the one subjected to the violence inflicted, it is proper to remark, that an eel may be cut
into several parts, and each give manifestations of life, in consequence of each part being possessed of one or more independent nervous ganglions. It will be perceived that a blow on *one
ganglion destroys life by its direct communication, by nervous
filaments, with the other ganglions,* whilst division of the eel into
parts communicates no direct violence to the ganglions the part
contains.

A drop of concentrated hydrocyanic acid placed on the
tongue will kill a rabbit; its destructive influence on the
organic nerves of the tongue is instantly propagated to the
organic ganglions and glands; vitality centred in the latter
is expelled, and death is the result. The shock is much the
same as when a blow is inflicted, and is instantly communicated to the whole organic nervous system.

Proofs to demonstrate that Life is not located in the Cerebro-Spinal System, inasmuch as Life exists where the Cerebro-Spinal System is wanting.

Mr. QUAIN says: "Now, as to the *sympathetic* nerve, so far from being in any way derived from the brain or spinal cord, it is produced independently of either, and *exists* notwithstanding the *absence of both.* It is found perfectly *formed* in *acephalous* infants, therefore *does not arise*, mediately or immediately, from the *brain;* neither can it be said to *receive roots* from the *spinal cord*, for it is known to exist *as early* in the fœtal state as *the cord* itself, and be fully *developed*, even though the latter is *altogether* wanting. It appears that whilst the organs of vegetation and life are being formed, the sympathetic nerves are *produced concurrently* with them; and that as the growth of these parts proceeds from the *circumference* to the centre of the whole body, from its lateral parts to the median line, the *sympathetic nerves* also *conform* to the general law."—See QUAIN's *Elements of Anatomy, p.* 711.

The class of animals known as the *Invertebrata* have *no cerebro-spinal nervous system,* and yet are fully *endowed* with *vital* powers.

The brain of a sheep or rabbit may be *totally destroyed* without *killing* the animal.* The head may be *separated* from the body by *decapitation* performed between the atlas and occipital bone, *without destroying* life, either in the *head* or *body.* The *mouth* of a sheep will *open* and *close*, and the *body* will give *evidence* of *strength* for some minutes. It will be remembered that although the brain and central ganglion has been destroyed, yet that the cerebral and spheno-palatine ganglions remain untouched; and that the muscles in the neighborhood of the mouth receive filaments from the spheno-palatine ganglion; as also that in the body, the prevertebral, the cardiac, the pulmonary, and semilunar ganglions are ganglions of the vital or organic nervous system, and continue untouched.

---

* As proved by vivisections by the author.

4

To prove the Identity of Animal Electricity given off by the Torpedo, and Electricity given off by the Pulmonary Glands.

The Pulmonary Organic Glands are capable of giving off Electricity on being Stimulated by the Entrance of Air into the Lungs, to unite the Oxygen mechanically mixed with Air with the Venous Blood, in its passage through the Pulmonary Glands.

"*Electricity—Animal.*—A power or imponderable agent possessed by and evolved from certain living animals, which enables them, independently of the operation of external agents on their structures, to produce several of the phenomena exhibited by common or voltaic electricity generated in organic matter."—JOHN COLDSTREAM, *Cyclopædia of Anatomy and Physiology*, vol. ii., page 81.

For the Evolution of Electricity, "there is no chemical action, no friction, no alterations of temperature, no pressure, no change of form. The exercise of the animal will, and the integrity of the nervous system, as well as of certain peculiar organs which exist in the animals endowed with electrical power, seem to be alone sufficient for it."—COLDSTREAM.

Description of Electrical Organs in the Torpedo.

Dr. COLDSTREAM says, "the partitions of the electrical columns are covered with a fine net-work of arteries, veins, and nerves."

Mr. HUNTER, in *Phil. Trans.*, 1773, page 481: "They are very vascular, sending inward from the circumference, all around on each partition, small arteries, which anastomose upon it, and passing also from one to the other, unite with the vessels of the adjacent partitions."

Electrical Organs amply furnished with Animal Nerves.

"The nerves of the electrical organs of the Gymnotus are derived from the spinal marrow alone. They are very large and numerous, and are divided into very fine twigs on the cells of the organ."—COLDSTREAM.

Electrical Organs can Evolve Electricity independently of Animal Nerves.

Mr. Todd (*Phil. Trans.*, 1816,) finds that *division* of nerves and *laceration* of electrical organs *did not prevent* the torpedo from giving an electrical shock.

Electricity more powerful in the Young than the Old Torpedo.

"The electrical power of the *young fish* is proportionally very much *greater* than that of the old, and can be *exerted without exhaustion* and *loss of life* much *more frequently.*"—Cold-stream.

Electricity Serves to Promote Respiration.

Dr. John Davy, (*Phil. Trans.*, 1835,) states that he thinks the electricity evolved when the torpedo is in mud or sand *assists* respiration, by *decomposing* the surrounding water and allowing the air to come in contact with the gills.

Overexertion of the Electrical Organ Destroys Life.

"All electrical fishes soon become exhausted and die, even in sea-water, when they are excited *to give a continual succession* of discharges."—Coldstream.

The *similitude* of a *section* of the *electrical organ* of the torpedo and a *section* of the *lung* is *very striking.* The sections of *both organs* are furnished with *arteries, veins, nerves,* and *cells.* In the torpedo, the electrical organs are amply supplied with animal nerves, derived from the spinal cord. In the human subject, the par vagi *(animal nerves)* are freely *distributed* to the lungs.

The torpedo *can evolve* electricity, even when the *animal nerves are destroyed. Respiration* can be carried on, as will be proved in another place, *when the vagi are divided.*

By repeated *shocks*, a torpedo will become *exhausted*, and either *die* or be *unable* to evolve electricity.

A man or a hare can be run to death in consequence of the respiration ceasing to be carried on, the pulmonary glands being unable to evolve electricity to unite the oxygen of the air with the venous blood in its passage through the pulmonary glands; death is caused in the same way as that by which repeated shocks kill the torpedo.

A young torpedo has a vastly greater power to sustain the shock given by the evolution of electricity than an old one.

A young man can withstand with impunity the shocks given when the semen is emitted, whilst an old man will quickly fall a victim to the shocks produced by sexual intercourse; precisely in the manner that repeated shocks will quickly kill an old torpedo.

A young torpedo is endowed with a vast amount of electricity, and will sustain life under circumstances that would *destroy vitality in an old one.*

A young child or infant has a greater amount of vitality than either a young or old man. A matter proved by the fact that an infant or child, on exposure to cold, will *survive*, when a young man or an old man will be *found dead* under similar circumstances.

Objections to the Theory that the Blood is Oxygenized by the Process of Endosmosis and Exosmosis.*

It is manifest, if the blood in the lungs were arterialized by the process of endosmosis and exosmosis, that establishing artificial respiration should prove successful in restoring and prolonging life. This theory, unfortunately, is not sustained by experiments.

There are certain diseases where the air enters the lungs, and where, if the blood were oxygenized by the process of endosmosis and exosmosis, no difficulty could be apprehended from the blood not being arterialized, but in which diseases there are ocular demonstrations afforded that the blood is not arterialized.

---

* This theory is the one promulgated by the most distinguished physiologists of the present day, and is so firmly believed to be true that no person has questioned its correctness.

Arguments to Sustain Objections.

The process of endosmosis and exosmosis implies an internal and external effort:(εν δος, inward; ωσμος, effort;)(εξω,outward; ωσμος, effort.) Wherefore, when no change can be detected in the organization of the parts said to be engaged in this process of the arterialization of the blood in the lungs, whether the blood is or is not arterialized, it is evident there must be other agents engaged in causing the blood to be arterialized, as well as to obstruct the process of arterialization, besides endosmose and exosmose.

Asthma.

To demonstrate that this proposition is founded on facts, it may be observed in a case of asthma, where the air passes into the lungs, that when a person enters a meadow, where the grass has been recently mown, he is liable to be attacked with difficult respiration and all the symptoms of spasmodic asthma. The same state of things may be induced by going into an apothecary-shop where hippo is kept. Again, a third person will actually faint on inhaling the vapors of a sweet-scented rose. The vapors from the hay, from the hippo, from the rose, all act in the same way. All pass into the lungs with the air; all come in contact with the pulmonary organic nervous glands, and interfere with or arrest its function of giving off electricity, to cause the union of the oxygen with the venous blood in passing through the glands. In case of spasmodic asthma, the countenance of the patient will be purple, the lips and tip of the nose particularly so, showing the absence of oxygen in the blood. In the case of fainting by the vapors of the rose, the pulmonary organic nervous glands have their function of giving off electricity momentarily suspended; hence fainting or suspended animation, for the want of oxygen to combine with the organic nervous glands, is the result.

Hooping-Cough.

When a child is attacked with hooping-cough, at certain intervals it will be seized with fits of coughing; the air will be continually expelled by forced expiration, until all is apparently forced out of the lungs. At the time, if the child's countenance is examined, it will be observed to be pale and livid. The patient now struggles for breath, and each inspiration is accompanied by a peculiar kink, until respiration is again established. Here the pulmonary organic glands are in a state of spasm; they become incapacitated to discharge their function of giving off electricity, to combine the oxygen with the venous blood; hence the pallid, livid countenance can be accounted for. The patient is on the verge of fainting, for the want of oxygen to combine with the organic nervous glands.

In the case just described, all air is forcibly expelled from the lungs. The muscles engaged in the process of respiration could never accomplish this end; the air is expelled by an imponderable force from the lungs; the pulmonary organic glands being in a state of morbid irritation, cannot bear the presence of the stimulus of the air. If the arterialization of the blood depended solely on the process of endosmose and exosmose, the mechanical condition of the lungs continuing the same, the arterialization of the blood should go on without intermission; but the state of the patient, as hooping-cough, shows the contrary is the case.

Morbus Cordis.

Another proof that the blood is not oxygenized in the lungs by the process of endosmose and exosmose, is furnished by the appearance and symptoms of a person laboring under great enlargement of the heart, with valvular disease. In a case of this kind, there is no obstruction to the air passing into the lungs by the windpipe; and consequently, if the blood were arterialized by the soaking in of the oxygen from the air, there could be no difficulty about the arterialization of the blood. But the congestion of the countenance; the lividity of the lips and nose; the distressing and suffocative respiration; the inability to remain in a recumbent position; the relief obtained by pressure of the chest against a chair or table; the laborious and confused action of the heart; the undulating and intermitting pulse; the cold extremities—all indicate great derangement of the organic nervous system. It is evident the blood is not sufficiently oxygenized. It is evident, therefore, the organic pulmonary glands are incapacitated from doing their duty. It is manifest the action of the heart is deranged. Here the difficulty is caused by pressure of the enlarged heart on the cardiac ganglion. When its anatomical relations are recollected, it will be at once perceived that the ganglion could not possibly escape pressure; and that the more it is pressed on, the more difficult the respiration becomes, until eventually respiration ceases altogether, in consequence of the pulmonary organic ganglions being rendered powerless by the continued pressure the cardiac or pulmonary ganglion is subjected to.

Vivisections

Made for the purpose of proving the Pineal Gland is a Ganglion of the Organic Nervous System, and of the same Class of Ganglions as the Lenticular.

In order to afford facility to others to make experiments for their own satisfaction, I will state the mode of proceeding. The best animal that can be selected for experimenting is a sheep; none other can be kept sufficiently quiet. The instruments required consist of a hand-saw, chisel, dissecting-knife, forceps and retractors, sponges, water; ligatures and plugs of paper, about a quarter of an inch in thickness, should be in readiness. The sheep, being placed on a firm table, should be held by assistants. The scalp is now to be freely removed, together with the muscles attached to the cranium. A piece of the skull is next to be taken away, about $3\frac{1}{2}$ inches long and $2\frac{1}{2}$ inches in width, transversely about one inch and a half above the superciliary ridges, extending posteriorly about one inch beyond the occipital protuberance. (See *Plates*, p. 62.)

In elevating the bone, the dura mater will be removed with it, leaving the cerebrum exposed to view. Separate the falx cerebelli from its attachments on each side; arrest the hæmorrhage, by promptly plugging the lateral sinuses. Having reached the fissure of Bichat, the posterior border of the corpus callosum, together with that portion of the fornix incorporated with it, must be divided in the mesial line from before backward, and held asunder by retractors; divide the velum interpositum in the same direction; arrest hæmorrhage by the application of cold water. The venæ Galleni, which carry blood from the plexus choroides to the straight sinus, cannot escape being divided. The tubercula quadragemina are now brought into view. A small, pale, yellowish-red body will be seen anteriorly resting against the nates, connected to the optic thalami at the sides, and placed just above the *iter a tertio ad quartum ventriculum;* this is the Pineal gland or ganglion. It will be perceived, in the latter part of the dissection, that every time the point of the knife goes into the neighborhood of the gland, the sheep plunges.

Experiments.

The gland being now open to observation, you gently seize its body with the forceps, extending the points towards or nearly as far as its attachment to the thalami. Oscillations of the iris will be at once the effect. Move the gland more freely, and the pupil will contract to a very small diameter in an instant. Still further press the gland, and make traction, and the eyeball will move rapidly in all directions. Make more firm pressure and traction, and the sheep will vomit and be thrown into a tetanic spasm; the neck will be curved, the legs thrust violently forward, every muscle in the body will appear to be engaged, and you will hear the by-standers exclaim, "You have killed the sheep!" The forceps being now relaxed, the sheep will shortly recover, and you can go through the same process. Sometimes, instead of the vomiting, the sheep will bleat most pitifully, and then be thrown into the tetanic spasm. Press the gland down towards its attachment, and you will observe the pupil to dilate.

First Sheep.—After exposing the brain, Dr. Busteed passed in a very fine needle, with a view of piercing the gland; the sheep was instantly attacked with violent convulsions, which continued until the butcher cut the sheep's throat. The brain was now removed, and the needle was found to have passed through the peduncle of the gland.

Second Sheep.—After exposing the gland, compression produced dilatation of the pupil and tetanic spasm, when the sheep's throat was cut.

THIRD SHEEP.—Hæmorrhage rendered the operation unsatisfactory.

FOURTH AND FIFTH SHEEP.—Contraction, dilatation, rolling of the eyeball, and tetanic spasms.

SIXTH SHEEP.—Similar results, together with vomiting, on the gland being withdrawn from its attachment, which was instantly followed by tetanic spasm and death of the sheep.

SEVENTH AND EIGHTH SHEEP.—Contraction and dilatation of pupil; rolling of the eyeballs. Pressure with traction caused the sheep to bleat most pitifully, as if suffering extreme torture, followed by tetanic spasms. In these two cases the experiment was performed by slicing off the lobes of the cerebrum on a level with the corpus callosum; reflecting the latter backward with the anterior pillars of the fornix, the sheep were observed to commence snoring during the operation. After the hemispheres of the brain were removed in both cases, the bodies of the lateral ventricles were slightly opened, allowing the blood to flow in. On removing the fornix and the clot of blood, the sheep breathed naturally. Supposing the seventh sheep had died in the tetanic spasm, the butcher cut off the head, when the body of the sheep plunged violently, and life was not extinct for about three minutes.

NINTH SHEEP.—Contraction and dilatation of pupils; rolling of the eyeballs; tetanic spasms and death. Here let me observe, the bleating of the sheep was caused by the irritation propagated through the recurrent branches of the par vagi, to the organic nerves in the larynx, derived from the superior cervical ganglion.*

Objection to SIR CHARLES BELL'S *Experiments on the Spinal Nerves, founded on* MR. GRAINGER'S *Description of the Prevertebral Ganglions.*

"On laying bare the roots of the spinal nerves, (says SIR CHARLES,) I found that I could cut across the posterior fasciculus of nerves, which took its origin from the posterior por-

---

* It cannot be denied but that it is a most cruel operation to perform vivisection, and nothing short of arriving at the truth of an important scientific investigation, I solemnly declare, would induce me to make one. •

tion of the spinal marrow, without convulsing the muscles of the back; but that, on touching the anterior fasciculus with the point of the knife, the muscles of the back were immediately convulsed."

Here, let me remark, the *anterior* and *posterior roots* of the *spinal nerves*, as Mr. GRAINGER has beautifully shown, *receive filaments* from the *prevertebral ganglions*, and *consequently* the anterior roots of the spinal nerves could not be *irritated* without *touching* the FILAMENTS of the GANGLIONS; thus *showing* the *fallacy* of the *experiment*.

Another Mode of Performing Vivisections,

Or another Mode of Experimenting on the Pineal Gland, to prove it to be a Ganglion of the Organic Nervous System.

Every operating surgeon knows that to perform an operation on the dead subject is quite a different thing to doing it on the living body.

The same remark is equally true with respect to exposing the Pineal gland in the living or dead animal. In the former there is great trouble; in the latter there is no difficulty. By a simple experiment, any person can satisfy himself that the Pineal gland is a ganglion of the organic nervous system, as piercing the gland with a fine needle will produce the following phenomena:

Oscillation of the iris, contraction of the iris, rolling and fixing of the eyeballs, and tetanic spasm of the muscles of the body.

It should be observed, unless the gland is touched, no such effects are produced by puncturing any other part; which fact affords in itself, in the strongest manner, negative proof of the importance of the ganglion.

A glance at the annexed plates, drawn by my friend, Mr. WILLIAM HENNESSY, from dissections made by myself, will show the mode of carrying out the vivisection, and demonstrate the relative anatomy of the ganglion.

If one line be drawn transversely, so as to allow the posterior lobes of the cerebrum to touch it, and another in the direction of the longitudinal fissure, the ganglion will be found in the median line, at a distance of three-quarters of an inch from the transverse, and at a depth of one inch and an eighth from the peripheral surface of the cerebrum.

In case an attempt is made to puncture the ganglion, and that it escapes being wounded, the sheep will fall into a state of coma, and commence snoring.

If the gland is now cut down on, it will be found surrounded by a clot of blood; on removing the clot, the respiration will become natural.

Dr. R. B. Todd's *Description of the Pineal Gland.*

"Pineal Gland.—We may here conveniently notice the position and connection of the Pineal gland. This body, rendered famous by the vague theory of Des Cartes, who viewed it as the chief source of nervous power, is placed just behind the third ventricle, resting in a superficial groove, which passes along the median line, between the corpora quadrigemina. It is heart-shaped, and of a gray color. Its apex is directed backward and downward, and its base forward and upward. A process of the deep layer of the velum interpositum envelops it, and serves to retain it in its place. From each angle of its base there passes off a band of white matter, which adheres to the inner surface of each optic thalamus. These processes serve to connect the Pineal body to the optic thalami. They are called the peduncles of the Pineal gland, also habenæ. In general they are two in number, one for each optic thalamus. They may be traced forward as far as the anterior pillars of the fornix. Posteriorly these processes are connected along the median line by some white fibres, which adhere to the base of the Pineal gland, as well as to the posterior commissure beneath, and which seem to form part of the system of fibres belonging to that commissure. A pair of small bands sometimes pass off from these fibres, along the optic thalami, parallel to the peduncles above described."

The Pituitary Gland demonstrated to be a Ganglion of the Organic Nervous System.

Nutrition, assimilation, secretion, and absorption are the result of organic nervous influence. How are these important matters provided for in the encephalon? Where is the organic ganglion to be found destined to preside over these functions in the brain? I will answer, In the sella turcica; and is that body called the pituitary gland, which is composed of gray and white matter, incased in the dura mater, enveloped by the arachnoid membrane, communicating through the infundibulum with the third, as well as all the other ventricles of the brain, and, by the continuity of surface of the arachnoid membrane, with the entire surface of the cerebral mass. I should also observe, that it appears to be in direct communication with the Pineal gland or central ganglion; that the pedunculi of the gland can be traced down towards the infundibulum. (See *Plates.*) It may be said no nerves can be detected in the arachnoid membrane; but the same objection holds good with respect to the pericardium, pleura, and peritoneum. However, when these membranes as well as the arachnoid are in a state of inflammation, the exquisite pain proves, beyond a doubt, the existence of nerves; the vascularity shows the presence of blood-vessels, although such could not be previously discovered; and the effusion of lymph, serum, or pus, demonstrates that secretion is vigorously carried on, and the subsequent removal of these substances points out the activity of the absorbents.

It will be perceived the semilunar ganglions perform in the abdomen the same kind of duties the pituitary gland or ganglion does in the cranium. The ganglions further resemble one another in being located in secure positions, being in the proximity of large blood-vessels; in being at some distance from the organs they supply with nerves.

The white bands constituting the pedunculi of the Pineal gland, as before stated, can be seen proceeding towards the infundibulum, which passes down from the third ventricle to the pituitary gland; the internal carotid arteries pass by the sides of the pituitary gland, surrounded by a plexus of nerves derived from the superior cervical ganglion; branches of the plexus enter the gland, so that a complete communication is established between all the ganglions in question.

FIG. 1.

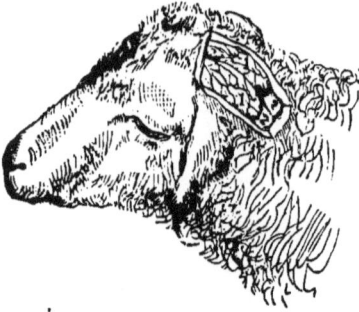

FIG. 2.

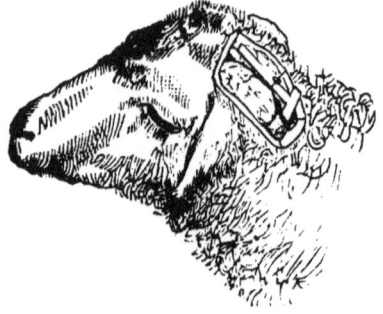

FIG. 4.

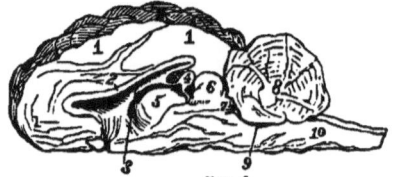

FIG. 3.

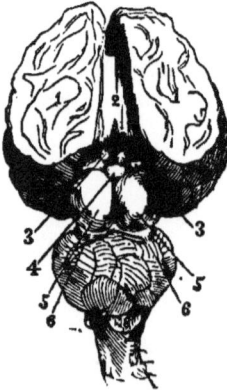

FIG. 5.

FIG. 6.

FIG. 7.

Fig. 1.—Appearance of Cerebrum and Cerebellum after the removal of the Calvarium and Dura Mater.

1, 1, Cerebrum.   2, 2, Cerebellum.

Fig. 2.—Similar View as Fig. 1, with rule and needle describing the method for finding the Pineal Gland or Ganglion.

Fig. 3.—Hemispheres of the Cerebrum removed on a level with the Corpus Callosum. The posterior lobes of the brain drawn upward and forward with the Corpus Callosum, so as to bring into view the Pineal Gland and its Peduncles.

1, 1, Cerebrum.   4, Pineal Gland.
2, 2, Corpus Callosum.   5, 5, Nates.
3, 3, Pedunculi of Pineal Gland.   6, 6, Testes.

Fig. 4.—Vertical Section of the Brain, showing the anatomical relations of the Pineal Gland.

1, 1, 1, Cerebrum.   6, Nates.
2, Corpus Callosum.   7, Testes.
3, Fornix.   8, Cerebellum.
4, Pineal Gland.   9, Fourth Ventricle.
5, Opticus Thalamus.   10, Medulla Oblongata.

Fig. 5.—View of the Base of the Brain, showing the position of the Infundibulum.

1, 1, Cerebrum.   7, Medulla Oblongata.
2, 2, Olfactory Nerves.   8, Fifth Pair of Nerves.
3, Optic Commissure.   9, Facial Nerve.
4, Infundibulum.   10, 10, Sixth Pair of Nerves.
5, 5, Third Pair of Nerves.   11, 11, Ninth Pair of Nerves.
6, Pons Varolii.   12, 12, Fourth Pair of Nerves.

Fig. 6.—Horizontal Section of the Brain.

1, 1, Cerebrum.   8, 8, Plexus Coroides.
2, Anterior Commissure.   9, 9, Nates.
3, 3, Corpus Striatum.   10, 10, Testes.
4, Infundibulum.   11, Valve Vieussens.
5, 5, Pedunculi of Pineal Gland.   12, Medulla Oblongata.
6, Pineal Gland.   A, A, A, Lateral Ventricles.
7, 7, Optici Thalamici.   B, Third Ventricle.

Fig. 7.—View of the Base of the Brain, showing the Connection of the Infundibulum with the Pituitary Gland, after its removal from the Sella Turcica.

1, Pituitary Gland.   2, Infundibulum.

Proofs that the Arteries are surrounded by Plexuses or Retinæ of Organic Nerves, which send Twigs into their Middle and Internal Coats, and accompany them to their Terminations.

Mr. HARRISON states *(Surgical Anatomy)* that the superior cervical ganglion sends branches in the carotid canal to the cavernous sinus, which *form* a plexus or *ganglion* on the external surface of the artery; that *fine, soft, reddish filaments* pass from the *cavernous* or *carotid* ganglion; that several are *attached* to the *carotid artery*, and *are lost* on *its cerebral branches*.

Mr. HARRISON says, "The inferior or descending branches of the superior cervical ganglion form a plexus *round* the *carotid artery*, from which *several branches* extend along the *external* carotid, *forming plexuses around* each of its principal branches, which are named accordingly."

Mr. HARRISON observes, that "several branches from the inferior cervical ganglion *encircle* the *subclavian* artery, and *extend* along its *trunk* and *its several branches*." He says, "these branches can be *traced* to a *great* extent on the *subclavian* and *axillary arteries*, FORMING PLEXUSES *in their* TISSUE."

Mr. HARRISON remarks: "A *considerable fasciculus ascends* along the *vertebral* artery, and *forms* plexuses around this vessel. These may be followed into the cranium; those of opposite sides unite on the BASILAR artery; they *follow* its *branches*, on which they *communicate* with analogous filaments from the carotid plexus."

Mr. HARRISON says, filaments from the superior cervical nerve *pass along* the coats of the *arteria innominata* to the aorta; at the arch of the aorta, some *filaments* pass *behind* and *before* it. Mr. HARRISON also states that the *aorta receives* branches from the *thoracic* ganglions, *as also the adjacent vessels*.

Mr. HARRISON remarks: "The *cardiac nerves* communicate with those of the other side, in the concavity of the arch of the aorta, both *below* and *above* the right or transverse branch of the *pulmonary artery*." He says: "The roots of the large vessels, and the structure of the heart, are *supplied* by *branches* from the *great cardiac ganglions* and *plexus*, and form the cardiac nerves." The *left* and *right coronary arteries are surrounded by plexuses of nerves, which accompany the arteries into the substance of the heart*, derived from the anterior and posterior coronary plexuses.

The *intercostal* branches of the aorta are *surrounded* by *plexuses* of nerves, *derived* from the *thoracic ganglions*.

Mr. HARRISON, in speaking of the semilunar ganglions, says: " This communication *surrounds* the *cœliac* axis, and the branches radiate to and from it in all directions; it is termed the solar plexus."

"From it numerous nerves pass off in various directions; those nerves *accompany* the *blood-vessels* and *form plexuses* around each, which are named, according to their destination, *hepatic, splanchnic, gastric, &c.*"

1st. The *phrenic plexuses accompany* the *phrenic arteries* to the diaphragm.

2d. The *supra-renal plexuses twine* round the *arteries* which accompany them to the *supra-renal* capsules.

3d. The coronary or *gastric plexus* accompanies the *arteria coronaria ventriculi.*

4th. The hepatic plexus—"its large posterior filaments *accompany the vena porta,* and its *anterior* the *hepatic* artery; these *accompany* the *vessels* in the *lesser* omentum to *the liver.*"

5th. " The *splenic* plexus proceeds in a similar manner *around* the *splenic* artery."

6th. The superior mesenteric plexus—" it *forms* a COMPLETE SHEATH for the superior *mesenteric* artery; its branches are numerous, very long, and distinct: they *accompany the arteries.*"

7th. " *The renal plexuses* are *formed* by *branches* from each side of the solar, *joined* by *lesser splanchnic* nerves; they *surround* the renal arteries,and accompany them into the kidneys."

In the male, each *renal plexus* gives off a fasciculus to *accompany* the spermatic artery, *around which it forms* the SPERMATIC PLEXUS, and *descends* to the testes.

In *the female, corresponding branches* from the *renal plexuses* supply *each ovary.*

5

8th. The inferior *mesenteric* plexus "*accompanies the inferior mesenteric artery* and its *branches.*"

9th. The *hæmorrhoidal plexus* is "continued AROUND the superior *hæmorrhoidal arteries.*"

The *abdominal aorta* is surrounded by nerves derived from the lumbar ganglions, between the superior and inferior mesenteric arteries; "the latter accompany the common iliac artery to their division, and *several filaments are prolonged around the internal and external iliac vessels.*"

The Hypogastric Plexus. All the Plexuses derived from this Plexus are conducted to their termination by the Arteries of each Organ.

"The arteries are plentifully supplied with nerves, of which the aortic system receives more in proportion than the pulmonary artery, and the *smaller arteries* more than the larger trunks. The trunk of the aorta, the pulmonary artery, and the arteries of the head, neck, thorax, abdomen, and those of the genital organs, *receive their supply from the nerves of organic life.* These form a very *intricate plexus* on their surface. *Two sets of nerves* have been *described* as being furnished to *the arteries*—one set *consisting* of *softer* nerves, of a *flattened* form, or said to be lost in the *cellular* or *external* tunic—*nervi molles;* the other set, more firm and round, PENE-TRATE THE MIDDLE *tunic*, in which they FORM a thin MEMBRANI-FORM expansion, *containing* DISTINCT *fibres.* MECKEL justly *considers* the *internal* nerves as SUBDIVISIONS of the larger *flattened external* BRANCHES."—JOHN HART, M.D., M.R.I.A.; Lecturer on Anatomy and Physiology, R.C.S., Ireland. *Cyclopædia Anat. and Phys.*, vol. i., p. 224.

Proofs from Comparative Anatomy that the Arteries are surrounded by a Retina of Organic Nerves.

Sympathetic System of the Cod.

"On each side of the aorta the prolongation of the sympathetic *is continued* down to the tail, *giving* filaments to the lateral *branches* proceeding from the aorta, and *communicating* with the *spinal* nerves; near the anus filaments are sent off, which *unite* and ACCOMPANY *the* SPERMATIC artery to the *ovaries.*"—T. RYMER JONES, Article "Pisces." *Cyclopædia Anat. and Phys.*, vol. iii., p. 998.

Extract from a Description of the Sympathetic System of the Boa Constrictor, by DR. SWAN.

"But at different points from which the nerves pass to the viscera, there is an appearance of a delicate plexus; this plexiform structure varies in different parts, and becomes much greater about the beginning of the intestine, where it resembles that corresponding with the semilunar ganglion in the turtle. Near the latter it assumes the form of a nervous MEMBRANE *or* RETINA, before it is distributed on the urinary and generative organs. BRANCHES pass from the PLEXUSES with the ARTERIES to the different viscera."

Pathological Proofs that the Arteries are surrounded by a Retina of Nerves.

In the first volume of the Transactions of the Physico-Medical Society of New York, for 1817, will be found a very valuable and ably written paper, entitled "Reflections on the Pulsations in Epigastrio, with an Inquiry into its Causes," by Valentine Mott, in which the following passages occur:

"That a pulsatory motion in the epigastric region should occur, unaccompanied with disease of any of the surrounding organs, is a *curious* and *interesting* fact. It is one of the most *extraordinary* and inexplicable phenomena attendant upon *nervous irritation.*"

Again: "That nervous irritation should here be *concentrated*, and develop itself in the form of a *pulsation*, is no more *extraordinary* than the phenomenon of BLUSHING."

Further on: "A very strong and regular pulsation was felt in epigastrio. It was so great, that MORGAGNI says he never saw it exceeded—it was very visible externally. The dissection of this patient showed no vestige of disease, either of the heart, large vessels, or abdominal viscera."

Professor MOTT's observation that the pulsation in the epigastric region is no more EXTRAORDINARY than the phenomenon of BLUSHING, IS UNDOUBTEDLY TRUE; the phenomenon in each *case depends* on *the action* of *the organic nerves surrounding the arteries.*

In the *London Lancet,* published in 1833, there is a case reported, which was under the care of Dr. WATSON, in the Middlesex Hospital, of a tumor in *the epigastric region,* which was *mistaken* by several practitioners, *who declared* it to be *aneurism,* and which *subsided* on the patient being *well purged.*

This case *confirms* the opinion *put forward* by Dr. MOTT, *that dilatation* of the *arteries* can be produced by *nervous irritation.*

When the tip of the index finger is attacked with *a whitlow,* the *radial* as well as *digital* arteries become *fuller, harder,* and *pulsate* violently; this condition of the arteries is *accompanied* by intense pain. The retina of organic nerves *surrounding* and *entering* into the *tissue* of the arteries, *continued* from the *axillary artery,* is in a state of *irritation;* hence the phenomena with respect to the condition of the arteries is susceptible of explanation, viz., "nervous irritation."

Positive and Circumstantial Proofs that the Capillary Arteries form Glands at their Terminations.

Now, it has been fully demonstrated that the *organic nerves surround* the arteries and *send branches* into *their tissue.*

It is fully ascertained that the blood in the smallest artery is *arterial.*

It is well known that the blood in the smallest vein is *dark-colored,* or *venous.*

It is a fact that the blood, at a point *corresponding* to the *termination* of the *capillary artery* and *commencement* of the *capillary vein, ceases* to be arterial, and *becomes* venous.

It is evident, therefore, the blood, in *passing* through the termination of the artery, *gives off its oxygen* and becomes *venous.*

It is a *chemical law,* that the *union of oxygen* with any other *matter* is attended with the *evolution of heat.*

It is a *physical law,* that the *production of heat is attended* with the *evolution of electricity.*

An excretory duct is characteristic of a secretory gland.

The salivary glands and the kidneys are examples of secreting glands: the former secrete the saliva, the latter the urine.

The kidneys are supplied with blood by the renal arteries, which are surrounded by the renal plexuses of nerves; the blood is removed from the kidneys by the renal or emulgent veins, and the urine by the excretory ducts of the kidneys and the ureters.

To test by direct experiment whether organic glands exist at the termination of the capillary arteries and the point where the capillary veins commence, let a man be told to run a mile, or until he gets fatigued: here it is to be observed, when a man runs, all the muscles are thrown into action; that the circulation is rapidly increased; that the venous blood is sent more rapidly by the action and pressure of the muscles to the right side of the heart, and from thence to the lungs; that the respiration becomes hurried; that more oxygen is combined with the blood; that burning heat of the surface is the result; that great thirst is complained of. Let a drink of cool tea be now given to him, and let him partake copiously of it, and witness what occurs. Almost immediately, the burning heat of the surface subsides, and an exhalation all over the body, from top to toe, will set in, followed by a copious perspiration from all the pores of the skin, as well as from the hair; (a hair is a hollow tube, and connected with a secreting gland.)

The explanation of what has taken place can now be expounded. The fluid taken into the stomach is rapidly absorbed and conveyed to the venous circulation; the blood, on being transmitted from the right side of the heart to the lungs, allays the excited condition of the pulmonary glands: and on being carried from the left side of the heart by the aorta and its branches all over the head, trunk, and extremities, the water supplied to the blood quenches the fire caused by the excess of oxygen introduced into the venous blood by the excited state of the pulmonary glands. The oxygen being in excess on uniting with the glands at the terminations of the arteries, increases the temperature of the surface to the highest degree. The *evolution* of *electricity* must *necessarily attend* the *evolution of heat*. The arterial blood containing the water. in its passage through the glands, has the latter decomposed by the electricity; the hydrogen of the water unites with the excess of oxygen in the blood, as well as some of the salts in the blood, forms water, which passes through the excretory duct of the gland, (the pore of the skin,) and thus soon removes the surplus oxygen from the blood.

Another Example to demonstrate the Pores of the Skin are in communication with the Organic Glands is well presented when a person is suffering extreme thirst.

Immersion in fresh water *quenches thirst ;* the water *passes* through the *pores* of the skin *into* the *glands ;* the *electricity* evolved on the union of the *oxygen* with the *glands* decomposes the water, the hydrogen of which *unites* with the *excess* of oxygen, *forms* water, which is carried *into* the venous circulation, and thus *allays* nervous excitement at the same time that it affords a certain amount of sustenance to the organs of the body.

The same explanation is true of persons who are immersed in salt water surviving, whilst persons on a raft perish.

The mode in which persons have the color of their skin changed by taking *nitrate of silver* for a length of time for the cure of epilepsy, *presents* a *beautiful* and *direct experiment* to *demonstrate* that the *oxygen unites* with the *organic* glands *situated* between the *capillary arteries* and *veins.*

When nitrate of silver is taken for a great length of time. the capillary nerves surrounding the capillary arteries become imbued with the *action* of the *nitrate* of silver, as well as the *organic* glands: the blood *circulating* through the *intestinal organic glands* is *next contaminated* or *impregnated* with the *silver:* the venous blood is carried from the glands thus charged with the silver to the right side of the heart, thence to the lungs. where it receives its oxygen, and next conveyed to the left side of the heart, from whence it is sent by the arteries all over the body; on the *union* of the oxygen with the *organic nervous glands.* the silver is also communicated to the glands. and in due time, when a sufficient quantity is *deposited* in the *organic nervous* glands, on *exposure* to light, forms AN OXIDE of *silver;* which accounts for THE COLOR of the skin, as already described.

From what has been just stated, the inference must be deduced, that there is a secreting organ or gland at the termination of the capillary artery and commencement of the capillary vein, inasmuch as that the blood *ceases* to be *arterial*, that the blood becomes *venous*, and that *secretion* also *takes* place; so that it becomes evident, without the intervention of a secerning organ, such changes could not be accomplished.

The organic gland is formed or composed of the termination of the capillary artery, the commencement of the capillary vein and the excretory duct, together with the organic nerves continued on the external coat and in the tissue of the artery.

Respiration.

The pulmonary ganglion sends a retina of nerves, which surround the pulmonary artery, sending twigs into its coats, and are thus continued on all the branches of the artery to their termination in capillaries, where they form organic nervous glands.

The *organic nervous glands*, therefore, are situated at the termination of the *pulmonary capillary arteries*, and give origin to the *pulmonary capillary veins*. The blood, therefore, has to pass through the gland before it can reach the vein.

The *glands* are in direct *communication* with the air-cells, which are *analogous* to the pores of the skin, which are in communication with the organic nervous glands of the skin. The organic glands in the lungs *communicate* with the air from within. In the skin, the organic glands are in communication with the air from *without;* hence the analogy between the skin and lungs.

As soon as the air comes in contact with the organic glands in the air-cells of the lungs, the glands are stimulated, and give off electricity, precisely in the same way as the electrical eel, when stimulated, gives off electricity, which causes the union of the oxygen of the air to unite with the venous blood which is passing through the gland, which, on being thus arterialized, is conveyed from the glands by the pulmonary veins to the left auricle of the heart.

On the union of the *oxygen* with the venous blood *heat is evolved*, and a certain amount of *electricity*, which *expels* the carbon and hydrogen, in the shape of vapor, from the lungs.

The Blood.

The blood is indispensably necessary for the support and continuance of life, and is *only second* in *importance* to the *organic nervous system*. It is the *medium* or *current* for conveying the *oxygen* to the *organic glands—it is, in truth, the stream of life;* besides, *it furnishes* the *materials* for the *regeneration* and *renovation* of *the various organs* of the *body*, under *the vital action* of the *organic nervous system*.

The Oxygen.

The *oxygen ranks* next in *importance* to the *blood*. By the *union* of the *oxygen* with the *organic glands*, the *operations* of *life are made manifest, characterized* by *respiration, circulation,* and *animal heat*.

The *cessation* of respiration for a short time is *attended* with *suspended animation* or *temporary death*. The total suspension of respiration is followed by death.

Vigorous respiration is accompanied by increased action of the heart and increased temperature of the surface of the body; whatever circumstance has a tendency to weaken or strengthen respiration is followed by a weakened or strengthened action of the heart, as well as a lower or higher temperature of the surface. Respiration being the effect consequent on the evolution of electricity or vital fluid by the pulmonary organic glands, it follows as a consequence, that whatever depresses or excites the organic nervous system weakens or strengthens respiration.*

---

* The experiments of Sir B. Brodie completely overthrow the doctrine of the chemists with respect to the production of animal heat; amongst them Liebig may be mentioned as the most celebrated.

Sir B. Brodie says, "that in an animal in which the brain has ceased to exercise its functions, although respiration continues to be performed, and the circulation of the blood is kept up to the natural standard, although the usual changes in the sensible qualities of the blood take place in the two capillary

To prove by demonstration the truth of the proposition
now propounded, it is necessary to give examples of cases
which occasionally present themselves:

First Example.

A man who receives a smart blow on the semilunar ganglion,
or the superior cervical ganglion, will fall to the ground; animation will be for some time suspended, in consequence of the
shock given to the ganglion being communicated to the pulmonary ganglion, as well as all the other ganglions, incapacitating the pulmonary organic glands to evolve electricity or
vital fluid to unite the oxygen of the air with the venous blood.

Second Example.

A delicate lady sometimes, on smelling a sweet-scented rose,
will faint; animation will be suspended for a longer or shorter
period. The vapor emitted from the rose passes with the air
into the lungs, and on coming in contact with the pulmonary
glands destroys their power of giving off electricity or vital
fluid to unite the oxygen of the air with the venous blood;
hence suspended animation or fainting ensues, in consequence
of the want of oxygen to unite with the organic glands.

Third Example.

A man falls from a height; he is taken up, apparently lifeless; his countenance is ghastly pale; his respiration is imperceptible; his pulse ceases to beat; his surface rapidly becomes cold; besides, he may have involuntary discharges.
The shock is communicated to the entire organic nervous system; the pulmonary glands are unable to give off electricity

---

systems, and the same quantity of carbonic acid is formed as under ordinary
circumstances, no heat is generated, and (in consequence of the cold air thrown
into the lungs) the animal cools more rapidly than one that is actually dead."—
Phil. Trans., 1811.

The shock the organic nervous system receives, when preparing to institute
the experiment, as performed by Sir B., accounts for the loss of temperature, as
will be explained more fully elsewhere. That PITIED or DECAPITATED ANIMALS
must sustain a tremendous shock, requires no arguments to *prove.*

or vital fluid to unite the oxygen of the air with the venous blood, in its passage through the glands: hence the suspension of life, for the want of oxygen to unite with the organic glands; respiration, therefore, as just stated, is imperceptible. The absence of the pulse is attributable to the loss of nervous power, produced by the shock; the coldness of the surface, or loss of animal heat, is caused by the want of oxygen to unite with the organic glands all over the body.

Fourth Example.

When a person has been immersed in water for some minutes, (three or four,) he will appear lifeless; all the operations of life will be suspended; yet life will still continue to occupy the organic nervous system, and may have its operations restored by proper treatment. The pulmonary glands, although for a moment they may be able to evolve electricity or vital fluid, yet have their functions quickly rendered powerless, and in a short time the operations of life cease all over the body, for the want of oxygen to unite with the organic glands.

Fifth Example.

When a man goes down into a deep well, that has been closed for some time, and which is full of carbonic gas at the bottom; or enters a tomb that has been recently opened, and is filled with sulphureted hydrogen, he is observed to fall suddenly. The instant the gases in question come in contact with the arytenoid muscles, spasm or closure of the glottis is the result: no air can enter into the lungs, consequently no oxygen can be supplied to the blood; and death necessarily follows, for the want of oxygen to unite with the organic glands.

The terms applied to the causes of death just mentioned, (in 4th and 5th examples,) in my opinion, are not philosophical. viz.: apnœa, ($\dot{a}$, priv., πνεω, spiro:) and asphyxia, ($\dot{a}$, priv.. σφιξη, pulsus.)

Sixth Example.

When a large quantity of blood flows from the body, fainting or suspended animation is the result. In such a case, the countenance becomes pale; the surface cold; the person tosses about, and gasps for air. The pulmonary glands are making strong efforts to sustain life, but are baffled in their attempts to do so, in consequence of the deficiency of the supply of blood to convey the oxygen to the organic glands: hence fainting or suspended animation is the result, caused by the want of oxygen to unite with the organic glands.

When all the blood is drained off, death is the result, in consequence of there being no blood to carry the oxygen to the organic nervous glands.

Seventh Example.

When a hare is run to death, the blood is found liquid and dark-colored. The organic nervous system has become completely exhausted; the pulmonary glands are unable to give off electricity or vital fluid to unite the oxygen with the venous blood: death is caused, therefore, by the want of oxygen to unite with the organic glands. (The appearance of the blood proves this statement to be true.)

Eighth Example.

When an angler has got a large trout on his hook, he gives him the full length of the line, allowing him to dart through the water freely, until ultimately, the fish, when completely exhausted, *floats* on the surface, and is easily captured. Here the organic nervous system has become *so exhausted* that the gills are no longer able to evolve electricity to *unite* the oxygen of the air contained in the water with the venous blood, and the fish *loses* the power of resistance.

Ninth Example.

Humboldt's description of the manner in which the *South American Indians* capture the *Gymnoti* is worthy of consideration in connection with this subject. A number of wild horses are driven into a pond in which the fishes inhabit. The fishes become excited by the presence of the horses; they make a furious attack on them; they give repeated shocks to the bellies of the horses, but they soon *become exhausted*, and float, almost lifeless, to the margin of the pool, when they are easily captured. The organic nervous system of the torpedo becomes exhausted on the same principle as the organic nervous system of the hare and the trout.

Certain agents increase the quantity of oxygen in the blood; others diminish it, by their action on the organic nervous system; whilst an operation may destroy its existence altogether.

A familiar example of the effect produced by the first agent is daily presented, viz.:

First Example—"Increase of Oxygen in the Blood."

A man drinks two tumblers of warm punch immediately after dinner; the punch stimulates and invigorates the organic nerves distributed on the internal surface of the stomach; the stimulation and invigoration are propagated to the whole organic nervous system; in addition, it may be stated, some of the punch is absorbed, passes by the thoracic duct into the venous circulation, stimulates the pulmonary glands in its passage through the lungs, as well as stimulates the organic glands on the union of the oxygen with the latter. Hence the glow of heat that pervades all the body; the flushed and animated countenance; the quickened respiration and great vascular excitement. To understand how all these matters have been brought about, it is necessary to recollect that the vital powers of the organic nervous system have been augmented by the action of the brandy; that the power of the organic pulmonary glands to give off electricity or vital fluid has been increased; that consequently a greater supply of oxygen is supplied to the blood, so that the quantity of oxygen being increased, the temperature of the surface of the body is elevated.

Second Example—"Decrease of Oxygen in the Blood."

Another man eats some tobacco; the tobacco depresses, nauseates, and exhausts the organic nerves of the stomach. The depression, nausea, and exhaustion are communicated to the organic nervous system; the man gets deadly sick; his muscular power rapidly diminishes; his heart pulsates weakly; his respiration grows feeble; his surface gets cold. To explain these symptoms, it is only necessary to remember that the effect of the tobacco on the organic nerves of the stomach extends to the whole organic nervous system, inasmuch as the whole organic nervous system is connected together; that the pulmonary organic glands, under such circumstances, are unable to evolve electricity or vital fluid to unite the oxygen with the venous blood; that consequently coldness of the surface must ensue, the quantity of oxygen in the blood being diminished; the feeble action of the heart and arteries is caused by the loss of vital power in the cardiac nerves, and the retina of nerves distributed to the coats of the arteries; the feeble respiration and tendency to faint are caused by the pulmonary glands being rendered incompetent to discharge their functions.

Third Example—" Cessation of Oxygen in the Blood."

When the external iliac or femoral artery is tied for aneurism, mortification is to be apprehended. The circulation of the blood being impeded, the supply of oxygen is cut off; the union, therefore, between the organic glands and the oxygen ceases; the provision for creating animal heat, by the union of the oxygen and organic glands, is suspended; coldness of the part is the result. The vital spark being extinguished for the want of oxygen, death takes place. The explanation of the causes which produce mortification of an extremity, where the circulation has been arrested, is now explained.

ANIMAL NERVOUS SYSTEM.

HAVING given a concise view of the organic nervous system of man, which I have demonstrated in another place is identical with the only nervous system to be found in the Invertebrata, I have to observe, that the animal or cerebro-spinal nervous system in man consists of the cerebrum, cerebellum, pons varolii, medulla oblongata, spinal cord, and cerebro-spinal nerves. The ANIMAL NERVOUS SYSTEM was given to man to enable him to *see*, to *hear*, to *taste*, to *smell*, to *touch*, to *think*, to *reason*, to *judge*, to *guide* the *movements of his body*, and to *express the sentiments of his mind*. The operations of the animal nervous system *are subservient* to the *will* and *guidance* of man, whereas he has *no control* over the *operations of the organic nervous system*. The *manifestations* of the *animal nervous system* depend on the *action* of the organic nervous system. It is therefore evident that the *animal nervous system* is of *secondary* importance, when *compared* with the *organic* nervous system.

Dr. TODD says: "The existence of this remarkable and peculiar kind of organic matter is LIMITED to the *animal kingdom*, and is therefore one of the *characteristic features* of *animals* as *distinguished from plants*. It is obviously the *presence of a psychical agent, controlling* and *directing* certain *bodily acts* of animals, which *has called* into existence the particular apparatus which nervous matter is employed to form."

Dr. TODD, further on, says: "The nervous matter is accumulated into masses, forming what are denominated centres of nervous actions; and it is also developed in the form of *fibres, filaments*, or *minute threads*, which, when bound together, constitute the nerves. The latter are *internuncial* in their office; they establish a communication between the nervous centres and the various parts of the body, and *vice versâ;* they conduct the impulses of the centres to the periphery, and carry the impressions made upon the peripheral nervous ramifications to the centres. Nor are the nerves mere passive instruments in the performance of their functions, but produce their proper effects through their susceptibility to undergo molecular changes, under the influence of appropriate stimuli."

The extracts just given from the late Dr. Todd's able and learned article on the nervous system AFFORD a sufficiently *explicit answer* to some *distinguished physiologists*, who *think* a *nervous system* is *not required:* or, in other words, that animals *could live* without such an *organization* as a *nervous system*.

ANIMAL LIFE

Is an intelligent agent or " imponderable power," located or centred in the animal nervous system, requiring intervals of repose, and requiring the agency of a physical agent, which is secreted by the cerebral organic glands, in order to make manifest its operations; as indicated by memory, judgment, reason, and volition.

Life, it will be recollected, located in the organic nervous system, requires oxygen, an external agent, to carry on its functions; whilst animal life, located in the cerebro-spinal system, only requires an agent furnished by the action of the organic nervous system; thus demonstrating that life, located in the organic nervous system, is incomparably of greater importance than animal life, located in the cerebro-spinal system; or, in other words, whatever destroys organic life must unquestionably destroy animal life, as the operations of the latter depend on the former.

The brain is largely and equally supplied with blood by the basilar artery, formed by the junction of the vertebral arteries and the internal carotids. The branches of these arteries form *a most remarkable anastomosis* or *inosculation*, called the "Circle of Willis." Mr. HARRISON says: "This is formed anteriorly by the *two anterior cerebral arteries*, with their cross uniting branch; laterally by each *internal carotid* and its *posterior communicating branch;* and posteriorly by the *trunk* of the *basilar* and the *roots* of the *posterior* cerebral arteries."

The *cerebral veins* or *sinuses* present the remarkable peculiarity of being *surrounded* by *strong fibrous* membranes, (duplicatures of the dura mater,) for the evident purpose of keeping off pressure, in case of distention from the brain.

The remarkable curvatures of the internal carotid arteries were made to impede the force of the circulation of the blood through them, and consequently prevent injury to the brain from this source.

The organic ganglions on the carotids in the cavernous sinuses, whose branches communicate with the nerves surrounding the basilar artery derived from the vertebral, regulate the circulation of the blood, and moderate its force in the circulation through the brain. The brain of a living sheep may be sliced off, and not a single artery will be perceived to give blood *per saltum;* thus showing the influence of the carotid ganglion, in regulating the force of the blood in the arteries distributed to the brain.*

Brain, (Seat of the Mind.)

The investigations of REID, GALL, and SPURZHEIM prove the *anatomical structure* of the *brain* to be FIBROUS.

Dr. TODD says: " *The principal bulk* of the hemispheres is formed by *fibrous structure;* this is shown by the *horizontal section which displays the centrum ovale.* These *fibres radiate from those surfaces* of the *optic thalami* and *corpora striata* which are in *contact* with the *substance* of the *hemisphere.*

" *Most of the fibres* which *emerge* from these *gangliform bodies pass to the gray matter of the convolutions;* some, however, *turn inward towards* the *mesial plane,* and *form* the *corpus callosum,* by *their union with those of the opposite side.*"

Nerve-Fibre.

Dr. TODD remarks: " REMAK and others describe three distinct parts in a nerve-fibre :

---

* In the elephant there is a *ganglion formed* on the *vertebral* artery.

" 1. *The outer investing membrane—tubular* membrane.

" 2. An inner layer of membrane (the WHITE SUBSTANCE of SCHWANN) being immediately within the first.

" 3. A central substance of nervous matter, called flattened band by REMAK, and supposed by him to consist of several filaments, or the axis cylinder of ROSENTHAL and PURKINJE."

Cerebro-Spinal Nerve.

"*A cerebro-spinal nerve consists of a congeries of fascicles, or bundles of nerve-fibres, or nerve-tubes enveloped and bound together by fibrous membrane, the nerve-sheath. The nerve-tubes, lie side by side, parallel, and sometimes have a wavy course within the general sheath.*"—TODD.

The size of the cerebro-spinal nerve-tubule in man is $\frac{1}{1833}$ to $\frac{1}{3300}$ of an inch.

Spinal Cord.

Dr. TODD remarks: " From a review of the preceding statements, it is plain that a large number of *fibres pass into the gray matter of the cord, and probably form some intimate connection with its minute elements; and the fact is favorable to the supposition that the spinal nerves derive their origin at least partly from the gray matter.*"

Gray Lines observed on Section of Spinal Cord.

STELLING and WALLACH suppose that these lines are *continuous* with the roots of the nerves; that they are, in fact, nerve-*tubes* proceeding from the gray matter to *form* these roots.

Dr. TODD says: "Microscopic investigation has as yet thrown no light on the direction and connection of the fibres of the cerebrum or cerebellum. What is known upon these points is derived from coarse dissection. The TUBULAR fibres of which the white matter is composed appear to be disposed on different planes, or, perhaps, interlace with each other, so as to render it difficult to isolate any plane to any great extent."

The arteries of the brain terminate in capillaries; the veins commence in capillaries; the cerebral organic glands are placed

6

between the extremities of the capillary arteries and the commencement of the capillary veins, in the gray substance of the brain and gray substance of the spinal cord; the *tubular* fibres of the brain are the *excretory* ducts, or receivers of the *secretions* of the cerebral glands.

During sleep the cerebral glands are in a quiescent state; but whilst awake are secreting a volatile agent, which stimulates the brain to action, so as to enable it to make manifest the operations of animal life, viz., reason, memory, and volition. In the *brain*, the volatile agent *passes* into the *tubules*, and is brought in contact with the white substance of the brain; in the spinal cord, the agent passes into the nerve-tubes at the extremities of the roots of the nerves. Thus it is that provision is not only made in the brain, but likewise in the nerves, for *receiving* the volatile agent secreted by the organic glands in the gray substance of the brain and spinal cord.

The necessity for having the nervous vesicular matter contained in nerve-fibres or tubules, and the necessity for the nerve-fibres or tubules to be incased in a strong covering or fibrous membrane, is obvious. The exceedingly delicate vesicular matter contained in the nerve-tubules would suffer disorganization, unless properly protected; the nerve-tubules being so very slender, would easily sustain injury, unless properly secured by the strong incasement of fibrous membrane.

Dr. TODD says: "The nerve-tubes lie side by side, parallel, and sometimes have a wavy course within the general sheath." "All observers, from FARLAND down to those of the present day, agree in denying the existence of any inosculation on anastomosis between the fibres in vertebrate animals; and it seems almost certain that this complete isolation of the nerve-tubes is not limited to those of the nerves, properly so called, but may be observed in the nervous centres also."

Each nerve-fibre represents some attribute of the mind; the interlacement of the trunks of the nerves in the brachial plexus allows the nerve-fibres to pass from one nerve to another; so that, by an exchange of the fibres of the various nerves entering the plexus, the operations of the mind can act harmoniously in the parts where the nerve-fibres are distributed.

Proofs to sustain the Theory that the Cerebral Glands secrete a Volatile Agent to stimulate the White Substance of the Brain contained in the Tubules and White Substance contained in Cerebro-Spinal Nerve-Tubes.

The results of vivisections, as given in another place, afford conclusive evidence that pressure by blood on the superior central ganglion will immediately induce sleep, or coma; the condition of the pupils, the circulation and respiration, as well as the loss of locomotion, denote that the organic nerves distributed to these organs are all suffering from the pressure inflicted on the superior central ganglion. It is evident the organic nervous system at all points is suffering from the disturbance caused by one of its principal ganglions sustaining injury. It therefore follows as a sequence, that the cerebral glands should also suffer, and have their functions arrested; thus it happens that the operations of the animal nervous system become suspended, as above described, for want of the stimulus of the volatile agent secreted by the cerebral glands.

Chloroform arrests the Action of the Cerebral Glands, and induces Sleep.

The vapor of the chloroform passes with the oxygen of the air into the arterial blood, and is communicated to the organic glands, on the union of the oxygen with the glands, narcotizing them; vital action is thus partially suspended; the cerebral glands have their functions arrested: non-secretion of the volatile agent is the result, and sleep sets in as a necessary consequence.

Before a person is put fully under the influence of chloroform, he gets apparently drunk: this takes place in consequence of the confused action of the brain, attendant on the irregular and interrupted secretion of the volatile agent by the cerebral glands; some of the glands being more under the influence of the chloroform than others.

Champagne excites the Action of the Cerebral Glands.

When a man has indulged in drinking Champagne or brandy punch, his whole organic nervous system will be thrown into a state of excitement, as indicated by the state of the circulation, respiration, and animal heat. The cerebral glands are also thrown into a state of excitement; a greater quantity of the volatile agent is secreted by the cerebral glands; the brain is stimulated; the operations of the mind are increased in activity; the person becomes loquacious and imaginative: fully verifying the doctrine of the poet—

Quem non fæcundi calices—desertum fecerunt.

Remarks.

In the case of chloroform, the whole organic nervous system is narcotized. As the operations of the mind cannot be carried on without the co-operation of the organic nervous system, it therefore follows, as a sequence, that the cessation of the cerebral glands to discharge their functions must be attended with the cessation of the operations of the brain. But, as regards the Champagne or brandy, the whole organic nervous system being stimulated and elated, the operations of the brain are hurried, as well as sometimes confused.*

Proof afforded by the Softening of the Brain, which occurs after tying the Common Carotid, that the Cerebral Glands discharge the Function of Secreting the Volatile Agent to act on the White Matter of the Brain.

It occasionally happens that after the carotid artery is tied,

---

* When a man is in a passion—when a man is excited by drinking Champagne—when a man's mind is elated by brilliant imaginations, his eyes *sparkle*. The explanation of the phenomenon consists in knowing that the cerebral glands secrete a volatile agent, on whose action the operations of the mind depend, through the *connection* of the *nerve-tubules* of the brain with the *cerebral* glands; and that phosphorus enters into the combination of the volatile agent, secreted by the glands, which passes into the *nerve-tubules* of the brain in the first instance, and the nerve-*tubules* of the *optic nerve* in the second instance, and manifests itself at the *termination* of the optic nerves in the *retina*, by the flashes of light which are *shot* forth from the eyes, under certain stages of mental excitement, or rather when the cerebral glands are vigorously engaged in discharging their functions. The appearance of the eyes indicates the presence of phosphorus; the condition of the urine demonstrates that phosphorus is secreted when the mind is actively employed; and further, it is to be remembered that the administration of phosphorus causes mental and nervous excitement. All these facts go to prove that phosphorus enters into the volatile agent secreted by the cerebral glands.

and that, too, in the most scientific manner, the patient will die, and that on post-mortem examination the side of the brain on which the artery has been tied will be found in a softened state.

Explanation.

The fatal occurrence cannot be attributed to a deficiency in the circulation of the blood in the brain, inasmuch as all the brain is *equally* supplied with blood, in consequence of the inosculation of the cerebral vessels in the "Circle of Willis." Therefore another cause must be found to explain the disorganization of the brain. The organic nerves surrounding the carotid artery sustain injury by the pressure of the ligature; the nerves become paralyzed; the paralysis extends along the trunk of the artery and its branches; the cerebral glands cease to act, the brain loses its usual stimulant, and softening of the brain is the result.

Reason, Judgment, and Memory are of a high or low Standard in MAN, in proportion to the Weight and Development of the Brain.

Proofs.

Tiedemann says, "The weight of the brain of an adult male European varies between 3 lbs., 2 oz., and 4 lbs., 6 ox.

"The brain of men with feeble intellectual powers is, on the contrary, often very small, particularly in congenital idiotismus. The brain of an idiot fifty years old weighed but 1 lb., 8 oz., 4 dr.; and that of another forty years of age weighed but 1 lb., 11 oz., 4 dr."

Dr. R. B. Todd says, "In all cases of idiocy there is a manifest imperfection in the development of the brain. This is sufficiently plain to the most superficial observer, from the small size of the head, which is so frequent a characteristic of this state, and which is more especially remarkable in adult life, when the development of the cranium by no means keeps pace with that of the rest of the body."

Whilst attending as a pupil at the Richmond Surgical Hospital, Dublin, I recollect Dr. R. Wm. Smith, the great pathological anatomist, (now Professor of Surgery at Trinity College, Dublin,) exhibiting the appearance of the brain of an idiot who manifested the lowest standard of intelligence, and who had recently died at the Richmond Lunatic Asylum. The cranium was very small and very thick, and the place which should be occupied by the anterior lobes of the brain contained a cyst filled with transparent fluid.

I have now, I think, sufficiently established the fact that animal intelligence, or the faculties of the mind, are located in the brain.*

The Brain of Man as compared with other Animals.

TIEDEMANN states: "SŒMMERING was the first to show that the human brain, in comparison to the *size* and *thickness* of the nerves, is larger than that of any other animal, even the elephant and whale, both of which have absolutely larger brains than man. It is also satisfactorily shown that the *organization* of the *human brain* is far superior to that of any other animal, not even excepting those apes which bear the closest resemblance to man."

The relative size of the brain when compared to the size of the animal, as well as the size and thickness of the cerebro-spinal nerves as compared with the brain, is well exemplified in the elephant.

Dr. R. B. TODD states: "The brain of an African elephant, seventeen years old, which was dissected by PENAULT, weighed 9 lbs. The brain of an Asiatic elephant weighed, according to ALLEN MOLINS, 10 lbs. SIR ASTLEY COOPER dissected an elephant's brain which weighed 8 lbs., 1 oz., 2 gr. (avoirdupois.")

It is scarcely necessary to observe, that the elephant is possessed of a vast amount of intelligence, and that such should be anticipated on remembering the weight and size of his brain, and might be supposed to possess higher intelligence even than man, had not SŒMMERING, TIEDEMANN, BLUMENBACH, OBEL, CUVIER, and TREVIRANUS satisfactorily explained the matter, as already stated.†

* CAMPER found the facial angle in the orang-outang to be 66; in the lowest tribe of the Hottentots, 70; in the Europeans, (Italians,) 76.

† Phosphorus becomes ignited or volatilized at the temperature of 100°—the

Connection of the Nervous Centres of Animal and Organic Life.

Reciprocity of communion is constantly taking place between the brain, the great animal nervous ganglion, and the superior central ganglion, which is one of the most important of the ganglions of the organic nervous system.

---

temperature of the blood at the heart is about 100°—just the temperature required to keep the volatile agent secreted by the brain in a state of activity. It will be remembered Mr. Hunter could never succeed in raising the temperature of an inflamed part above the temperature of the blood of the heart. In some diseases the temperature is raised to 106° to 107°, but it is a remarkable and curious fact that the temperature does not reach 108°, at which point phosphorus melts. Such a contingency might be attended with the very rare fatality known as "spontaneous combustion." In inflammation, the blood is increased in temperature by the phosphorus in the serum of the blood; in arachnitis the pain is excruciating, whilst the patient is furiously delirious in consequence of the *overstimulation* of the nerve-tubules of the brain by the volatile agent, which is increased in strength. The part the volatile agent plays in the eyes of wild animals is curious and remarkable, furnishing these animals with lamps to examine the pictures of objects impinged on the retina. The phosphoric volatile agent secreted by the cerebral glands, and transmitted through the nerve-tubules, illuminates the interior of the globe of the eye, so that the eyes appear of a dark night like balls of fire.

Any person can satisfy himself of the presence of the phosphoric agent in the eye, by closing his eye in the dark, and pressing on the ball at its inner angle, when he will perceive a luminous spot at the outer angle, the reflection of the phosphoric agent from the retina. Another proof that it is necessary to have the temperature of the blood up to 100°, to keep the volatile agent in a state of activity, is afforded by the state of the fingers when exposed to cold. The will commands certain movements of the fingers, but the fingers cannot carry the operations of the mind into execution, in consequence of the want of power, caused by the depression of the volatile agent. In cases of paralysis, the limbs affected are colder than the sound parts of the body; the volatile agent is not transmitted through the nerve-tubules of the nerves in these parts, hence the diminished temperature. The union of the oxygen with the organic glands keeps up a certain amount of heat, which heat is still further augmented by the transmission through the animal nerve-tubes of the volatile phosphoric agent. The temperature of the body is lower during sleep, when the cerebral glands are at rest.

" Phosphorus ($\phi\omega\varsigma$, $\phi\epsilon\rho\omega$, I bring light,) exists in Nature, *principally in the animal kingdom*; in the bones of the vertebrated animals; in the *fluids* of the body; and also in the *pulpy material* of the brain and nerves."—*Kane's Chemistry.* p. 295.

Phosphorus.—" It takes fire at 100°; melts at 108°."—*United States Dispensatory.* p. 556.

"Grief."

When grief or anxiety harasses the mind, the organic nervous system sympathizes and fully participates in the troubles of the mind. The painful sensation and oppression experienced about the heart, in popular language, is the result of the communication established between the superior central ganglion, the cardiac ganglion, and the semilunar ganglions, through the branches of the par vagi, (animal nerves,) which inosculate with the branches of the solar plexus and the branches of the cardiac plexus.

Connection between the Brain, the Superior Central Ganglion, and the Lenticular Ganglion.

When a person looks at a dazzling or luminous body, such as the sun, the iris or pupil instantly contracts, in order to exclude the rays of light as much as possible, and thus protect the retina from injury. The retina is the termination or expansion of the optic nerve, (an animal nerve;) the mind becomes cognizant of the danger which threatens it, but has not the power within itself to ward off the danger; it therefore communes with one of the vital ganglions, (the superior central,) whose duty it is to protect from injury as well as keep in a state of preservation the various organs of the body. The superior central ganglion communes with the lenticular through a *very small* branch of the third nerve, (an animal nerve.) The iris is supplied with nerves from the lenticular ganglion, and it at once contracts under their influence, and excludes the rays of light as much as possible.*

The mind listens to various sounds through the *portio mollis,* (an animal nerve,) distributed to the internal ear; but when it becomes necessary, or it is the wish of the individual to hear a sound at a great distance, the mind communes with the superior central ganglion; the latter affords the assistance re-

---

* It is to be remarked, the lenticular body is such an extremely small body, that it takes an expert anatomist to find it. Notwithstanding which, it controls the movement of the iris, causing its contraction or dilatation to meet the requirements of the mind.

quired, by communicating with the otic ganglion through a branch of the facial nerve; the tensor tympani muscle, being supplied with nerves from the otic ganglion, at once contracts, rendering tense the drum of the ear, and thus affording the facility to the mind it required for listening to the distant sound.*

"*Frightful Intelligence.*"

When a man is told disastrous news, the mind, located in the brain, communes with the superior central ganglion, (a vital organ;) the trouble is communicated to the cardiac ganglion, to the pulmonary ganglion, to the semilunar ganglions, to the renal ganglions, and spermatic ganglions, by the par vagi and its branches, inducing palpitation, suspended animation, or fainting, loss of appetite, diarrhœa, involuntary discharge of urine, and in the event of the person being a female, three months or eight months pregnant, abortion or death of the fœtus. In the case just detailed, the whole organic nervous system is involved in the trouble, and in the case of a female the trouble is communicated to the fœtus in utero, through the retina of organic nerves surrounding the maternal and fœtal vessels, which inosculate in the placental lobule.†

"*Playing on a Musical Instrument.*"

A person playing on a musical instrument affords a good illustration of the connection and communion between the animal and organic nervous systems. The mind wills that the finger shall touch certain notes in rapid succession. Certain

---

* It will be remembered that the tensor tympani muscle, acting under the influence of the otic ganglion, contracts or relaxes to meet the requirements of the mind with regard to hearing sounds at long or short distances.

† " For behold, as soon as the voice of thy salutation sounded in my ears, the infant in my womb leaped for joy."—*St. Luke*, i : 44.

" And it came to pass, whensoever the strange cattle did conceive, that Jacob laid the rods before the eyes of the cattle in the gutters, that they might conceive among the rods."—*Genesis*, xxx : 41.

Jacob was inspired; otherwise it is not likely he would have hit upon so true a physiological experiment. Any person can verify the truth of Jacob's experiments who has some *white* rabbits.

It is a remarkable fact, that the Deity has invariably carried his Omnipotent power into operation through the agency of *material* agents; a fact which will be at once admitted by any one who has read the Old and New Testaments.

relaxations and contractions of the muscles are required to produce the effects contemplated by the mind; although the mind wills the movement of the fingers, yet it has not the control over the muscles to insure their proper relaxations and contractions to meet the requirements of the mind. The mind communes with the superior central ganglion; the latter with the ganglions at the roots of the nerves which go to form the brachial plexus; the animal nerves thus placed in communion with the organic ganglions, after forming the brachial plexus, are distributed to the muscles of the forearm and hand, where their capillary or terminating branches inosculate with the organic nerves on the coats of the arteries. The organic nerves, in company with the arteries they surround, are distributed to the muscular fibres, and cause their contractions and relaxations to meet the requirements of the mind.*

"Singing."

When a singer wishes to produce certain notes, the mind, through the recurrent branches of the par vagi, instructs the larynx; the latter is supplied with numerous muscles to render lax or tense the vocal cords; but the mind having no control over the muscles, so as to produce their relaxation or contraction at pleasure, communes with the superior central ganglion; the latter communes with the eighth pair of nerves in the basilar plexus, and through these with the superior cervical ganglions; as the latter send retinæ of organic nerves round the arteries to the muscles of the larynx, the muscles contract and relax to meet the requirements of the mind, and bring out the required note by the vocal cords.†

---

* The absolute necessity for having a great number of nerve-fibres inclosed in the sheath of a single nerve is now quite evident, particularly when the numerous and varied operations of the mind, as exemplified in playing a musical instrument, are recollected—as the action of the fingers represents the operations of the mind.

† The brain represents the animal nervous system. The superior cervical ganglion represents the organic nervous system. The musical note represents the combined wisdom of the animal and organic nervous systems. Without being deemed, I trust, too *fanatical* or *presumptuous*, I beg to state that a very strong analogy is presented between the TRINITY of PERSONS in the DEITY and

"Leaping."

When a man wishes to leap a certain distance, for instance, from the top of one high place to the top of another high place, where the distance must be measured exactly in the mind to prevent falling, the mind wills the act, but has no control over the action of the muscles as regards their proper relaxation and contraction. The mind communes with the superior central ganglion; the latter with the prevertebral ganglions, at the roots of the nerves; the spinal nerves are distributed to the muscles; their terminating branches inosculate with the organic nerves on the coats of the arteries; which latter, being distributed to the muscular fibres, produce the precise condition required by the mind.

"Looking behind the Shoulder."

When a person wishes to look behind his shoulders, the will directs the act, but has no power over the muscles, so as to be able to make them contract harmoniously together, to meet the requirements of the mind. The mind communes with the superior central ganglion; the latter communes with the superior cervical ganglion, which latter regulates the proper axis of the pupil. through its connection with the lenticular ganglion; as well as the proper abduction of the eye, through its connection with the sixth nerve in the cavernous sinus, (which is distributed to the external rectus muscle;) and also the muscles of the neck, through its connection with the roots of the spinal nerves.

"Dislocation of Shoulder."

In a recent dislocation of the shoulder, the surgeon places his index and middle fingers of one hand under the head of the humerus, (if the dislocation is in the axilla,) and grasping the elbow of the patient with the other, he directs the attention of the patient to some extraordinary subject, and at the

---

the attributes of man: *"Let us make man to our own image and likeness."*— Genesis, i: 26. The organic nervous system represents the FATHER. The animal nervous system, the SON. The melodious note, the HOLY GHOST. I suggest this explanation for the benefit of disbelievers in the TRINITY.

same moment, by well-directed action of both hands, reduces the dislocation. Thus, by throwing the patient off his guard, he has no difficulty in reducing the dislocated bone. In the event, however, of his not being able to distract the patient's attention, this easy method may fail, inasmuch as, so long as the patient keeps his eye upon the affected shoulder, the animal and organic nerves will act in concert, and by strong contraction of the muscles, baffle his attempt. It will be observed, no matter how willing the person is to assist the surgeon, he cannot do so. The superior central ganglion being a vital ganglion, knows that on the reduction of the dislocation a sudden shock would be given, and guards against the occurrence, through its communion with the prevertebral ganglions, at the roots of the several nerves which go to form the brachial plexus, and from which branches are distributed to the muscles of the shoulder, terminating in filaments which inosculate with the organic nerves on the coats of the capillary arteries, which are distributed to the muscular fibres; hence spasm or rigid contraction of the muscles follows.

Vital Actions—(Excitation of Organic Nerves.)

Connection of the animal and organic nerves, as may be observed in the secretion of saliva, the secretion of milk, the secretion of gastric juice.

"Secretion of Saliva."

When a person longs for food, the popular expression is, that his "teeth water," or a copious secretion of saliva takes place.

Explanation.

The brain, the animal nervous ganglion, communes with the superior central ganglion; the latter communes through the brain and superior maxillary nerve (the second division of the fifth) with the spheno-palatine ganglion; the spheno-palatine ganglion communicates with the submaxillary gland, through the chorda tympani nerve, which is distributed to the gland, and inosculates with the organic nerves surrounding the capillary arteries; the capillary arteries become dilated through the organic nervous retina; more oxygen, as well as blood, is admitted into the arteries, and more heat, as well as electricity, is evolved. The electricity decomposes the serum of the blood, the hydrogen of the latter unites with the oxygen, and some of the salts of the blood secreted by the gland forms saliva, which passes off by the excretory ducts of the gland.

"Bashfulness."

A good illustration of the connection between the animal and organic nervous systems is afforded by watching the appearance of a bashful person, when charged with some ludicrous act.

The brain communicates with the superior central ganglion, and the latter communicates, through the brain and facial nerve, with the organic nerves surrounding the transverse facial artery, (it will be recollected branches of the facial nerve accompany this artery;) the result is, that the capillary arteries at once become dilated, and are injected with arterial blood, giving a crimson color to the face, at the same time that the temperature of the cheek is increased, in consequence of the large quantity of oxygen supplied to the organic glands.

"Grief."

When a person is afflicted with grief, the brain, the chief ganglion of the animal nervous system, communes with the superior central ganglion, the chief ganglion of the organic nervous system. The latter communes, through the brain and lachrymal branch of the ophthalmic division of the fifth nerve, (animal,) with the organic nerves surrounding the capillary arteries, and entering into the formation of the organic glands, in the lachrymal gland.

Explanation.

When the distribution of the nerve is recollected, as well as that it inosculates with the organic nerves of the parts to which the branches are distributed, there cannot be much trouble in accounting for the suffused condition or redness of the eyes, the capillary arteries being dilated so as to admit blood where none was visible previously. The mode in which the tears are secreted by the organic glands can be explained by recollecting the capillary arteries are dilated, carry more blood, as well as oxygen, and consequently, that on the union of the oxygen with the glands, more heat is the result, as well as a greater amount of electricity: that the electricity decomposes the serum of the blood in its passage through the gland, and that the hydrogen unites with the oxygen, as well as some of the salts of blood secreted by the gland, forms tears, which pass off by the excretory ducts of the gland.

Milk-Draught.

A familiar example of the connection between the animal and organic nervous systems is presented in what is commonly called the milk-draught. As soon as the mother sees her child after a short absence, her breasts will instantly become filled with milk. Here the explanation consists in recollecting that the brain communes with the superior central ganglion; the latter, through the spinal cord, with the *inferior cervical* ganglion, which communicates through the thoracic branches of the brachial plexus with the organic nerves surrounding the branches of the mammary artery; dilatation of the arteries takes place; more blood enters the arteries, consequently more oxygen. The organic glands commence their operations, and secrete the constituents of the milk; the electricity evolved on the union of the oxygen with the glands decomposes the serum of the blood; the hydrogen of the latter unites with the constituents of the milk, combined with oxygen, and forms milk.

Mental Emotions.

Mental emotions of an amatory character demonstrate the connection between the animal and organic nervous systems.

The excitement, being of a pleasurable description, is communicated from the cerebrum, the great animal nervous ganglion, to the superior central ganglion, the great chief of the organic ganglionic system; the latter communes through the brain and facial nerves which inosculate with the capillary retina of nerves surrounding the coronary arteries of the lip; the result is, the arteries become dilated, and contain more blood; hence the pouting of the lips that ensues.

Again, further communication is had through the thoracic branches of the brachial plexus which inosculate with the organic nerves surrounding the mammary arteries distributed to the nipple; the arteries become dilated and injected with blood; hence the erection of the nipples is the consequence.

Again, through the pudic nerve, whose branches inosculate with the organic branches of nerves surrounding the pudic artery, and distributed to the clitoris; hence dilatation, injection, and erection of the clitoris, and, in the male subject, erection of the penis.

Vital Action—Depression of the Organic Nervous System.

Why a person who is told dreadful intelligence gets deadly pale, is accounted for by recollecting that the brain communes with the superior central ganglion, and the latter through the brain and filaments of the facial nerve, which inosculate with the organic nerves surrounding the transverse facial arteries, rendering the nerves powerless, causing contraction of the capillaries; the circulation becomes feeble and the countenance pallid.

" Vomiting."

When a person, whilst at breakfast or dinner, is told a disgusting story, or happens to see a disgusting object, his stomach rejects its contents. The brain communes with the superior central ganglion; the latter through the branches of the par vagi, which inosculate with the organic nerves of the stomach surrounding the arteries; the nerves at once contract, followed by contraction of the capillary arteries, as well as by contraction of the muscular fibres of the stomach, to which the arteries are distributed. Contraction of the whole stomach is the result, with the discharge of its contents.

" Sea-Sickness."

When a person is standing on the deck of a ship which is tossed about by waves in a storm, the various objects are impinged on the retina so rapidly that the mind has not time to take cognizance of them, and becomes confused and disturbed; the confusion and disturbance are propagated to the superior central ganglion, and through the branches of the par vagi to the organic nerves surrounding the arteries of the stomach, with which the branches of the par vagi inosculate; the confusion and disturbance of the mind are thus communicated to them, causing contraction of the nerves surrounding the capillary arteries. Contraction of the arteries must be consequent on the contraction of the nerves, and contraction of the muscular fibres of the stomach must be consequent on the contraction of the arteries. Contraction of the stomach, rejection of its contents, and dreadful nausea, are the results of the disturbance and confusion propagated by the animal to the organic nervous system.

Depression, Excitation, Irritation of the Organic Nervous System, produced by External Agents.

Having demonstrated that the organic nervous system could be depressed or excited by its connection with the animal nervous system, I will now demonstrate that the organic nervous system can be depressed, excited, and stimulated by external agents.

Pneumonia.

When a person is exposed to rain and cold for the greater part of the day, he is liable to be seized with great depression, coldness of the surface, and shivering; the organic nervous system is greatly depressed, and its functions impeded. After a certain interval, the surface grows warm and hot, the countenance becomes flushed, and the pulse strong and full. Reaction has taken place, and the organic nervous system is in a state of high excitation. After a short time, heavy pain will be felt in the side, accompanied by oppressed respiration. The excitation of the organic nervous system is now followed by irritation, which accounts for the pain in the side.

Phlegmon.

Constant exposure to wet and fatigue is very often followed by an abscess, or what is usually called a phlegmon. The patient, after experiencing chilliness for some time, has his attention attracted to a certain part of the body or extremities, which has become hot and swollen. After the lapse of some time, severe pain, of a throbbing character, is felt in the part. The chilliness denotes the depressed state of the organic nerves; the heat affords evidence that the organic nerves surrounding the arteries have recovered from the shock; that dilatation of the capillary arteries has taken place; that an increased quantity of blood, and consequently, an increased quantity of oxygen, flows into the capillaries, and consequently that the temperature of the part is increased, on account of the larger amount of oxygen supplied to the organic glands. The pain demonstrates that the organic nerves are suffering from irritation, and are about throwing off the offending matter by the secretion of lymph and pus. The phlegmon, it will be perceived, affords a good example of the depression, excitation, and irritation of the organic nervous system; besides, gives an excellent illustration of the phenomena of what is generally or universally called inflammation.

" Rubor et tumor cum calore et dolore."

Excitation and Irritation of the Organic Nerves on the External Parts of the Body.

" Chancre."

The application of syphilitic poison to the organic glands of the prepuce or glans penis is not followed by any change in the part for some days. Itching and redness of the part will first attract attention; heat and pain immediately follow. A conical elevation (papilla) surmounted with a vesicle, to be replaced with pus, is lastly observed.

The poison takes some time to imbue the organic glands with the virus; consequently, during the incubation of the poison, no change is observable; but as soon as the poison has taken effect, excitation of the nerves takes place, followed by dilatation of the arteries and a larger flow of arterial blood into the arteries, and consequently an increase of the temperature of the part, by the additional supply of oxygen furnished to the organic glands. The pain is caused by the irritation consequent on the continued excitation of the nerves, and gives

7

evidence that the organic glands are about attempting to expel the poison with which they are contaminated, by the effusion of lymph or pus.

"Vaccine Pustule."

In about four days after vaccine matter is applied to the organic nerves of the arm, by making small incisions on the skin, redness of the part will be observed, and on the eighth day a vesicle (ultimately to be followed by a pustule) will be formed. The poison having occupied four days in impregnating the glands with its virus, no change takes place in the part to attract attention during that period; but the glands being now charged with the poison, causes their excitation, and consequently, dilatation of the arteries, allowing a larger quantity of blood and oxygen to enter them; hence the increase of temperature, as well as the effusion of lymph and serum, which follows, as just explained in reference to chancre.

Poisoning of the Organic Nervous System in the Interior of the Body by External Agents contained in the Atmosphere.

Depression, Excitation, and Elimination of Poison, by Organic Glands.

"Small-Pox."

When a person remains for some time in a room where a patient is laboring under small-pox, he is liable, (if not previously vaccinated,) after some days, to be attacked with a violent chill, to be afterwards followed by great excitement, characterized by heat of skin, bounding pulse, thirst, headache, great pain in the back, as well as severe muscular pain in the extremities. The patient will be in what is termed, in popular language, a high fever; and sometimes physicians have mistaken it for such. Between the fourth and fifth days the face and upper extremities will be remarked covered with pimples, to be followed on the ninth day with pustules, which become mature about the twelfth or thirteenth day, and which present the form of crusts or scabs on the eighteenth.

Explanation of the Phenomena.

The poison which generates small-pox, as in the case just described, is of an immaterial character; that is to say, it cannot be analyzed or measured by any chemical contrivance. It is mechanically mixed with the atmosphere, and passes into the blood with the oxygen; and is thus communicated to the organic glands all over the body, on the union of the oxygen with the organic glands. The poison, as in the case of syphilis, takes a certain time to bring the glands under its full influence; but, as soon as they are completely contaminated, they experience a great shock; the portals of life are shaken to their foundation, as made manifest by the severe rigor. The organic nervous system, true to its instincts, rouses into action, to expel the enemy which threatens its existence, and hence the cause of the excitement which follows. The appearance of the pimples, followed by the pustules, shows that the organic nervous system has commenced operations to clear off the poison, in the shape of pus, which it usually accomplishes in about eighteen days, provided the vital power in the organic nervous system is able to exist so long.

Depression, Excitation, Partial Loss of Vitality of Organic Nervous System.

" *Typhus Fever.*"

When a man resides in a neighborhood where typhus fever is prevalent, he is placed in a position to be attacked at any moment with this formidable disease. Generally, loss of appetite, debility, and a certain degree of uncomfortableness trouble the patient for a few days before the marked rigor sets in, which announces that the organic nervous system is fully under the impression of the poison. The rigor is not so strong in typhus fever as in small-pox, nor is the excitement of the organic nervous system so prominently marked as in the former disease. Debility, in aged persons, or in persons of vitiated constitutions, sets in very often on the ninth day, and the patient dies on the eleventh or twelfth. Between the fifth and twelfth days the patient's body will be observed to present a maculated appearance, being covered with small spots, at first red, but which gradually grow darker and darker, in proportion to the malignancy of the fever.

Explanation of the Phenomena.

The poison of typhus fever being immaterial in its character, is mixed with the air, and passes with the oxygen of the air into the blood, and poisons the whole organic nervous system, on the union of the oxygen with the organic glands. The indisposition complained of for a few days previous to the attack is indicative that the organic nervous system is feeling the injurious action of the poison; the rigor establishes the fact that the organic nervous system is fully under the impression of the poison; the reaction denotes the efforts of the organic nervous system to recover from the shock. The maculæ indicate the action of the poison on the organic glands, and an abortive effort to clear it off; the poison requires generally from fifteen to twenty-one days for its elimination.

Depression, Excitation, Secretion of "Organic Glands."

"Intermittent Fever."

A person residing in what is usually called a malarious district is generally attacked with what is commonly called fever and ague, or intermittent fever. When the disease is fully established, the patient is attacked with a severe rigor, accompanied with great coldness of the surface, and gnashing of the teeth; the organic nervous system is greatly depressed. After the lapse of a certain time, variable in duration, excitation of the organic nervous system takes place, characterized by great vascular excitement, flushed countenance, bounding pulse, burning heat of surface, urgent thirst, headache, and muscular pains. After the fever has lasted some time, the organic glands throw off the poison, by the process of secretion or transudation, through the pores of the skin.

Explanation.

The malarious poison is immaterial; it cannot be examined by the chemist; it commingles with the air, passes with the oxygen into the blood, and poisons the organic glands, on the union of the oxygen with the latter. The organic glands become imbued with the malarious poison, and give evidence of this fact by the rigor and depression of the organic nervous system which follow. The reaction shows that the organic nervous system has recovered from the shock, and is about to discharge the poison which harasses it, by the secretion of the organic glands, or perspiration through the pores of the skin.

It is the function of the organic glands of the skin to secrete or remove detrimental matters which are injurious to them: precisely in the same manner as the organic glands secrete the urine in the kidneys, or the organic glands secrete the bile in the liver.

Poisoning of the Organic Nervous System by Poison located in a certain Part of the Body through the Agency of the Blood.

" Secondary Syphilis."

When a man gets a true or Hunterian chancre, and succeeds in getting the ulcer healed by local applications and constitutional treatment, but leaving the *site* of the chancre remarkable for hardness, it is to be expected that after a certain period he will be attacked with what are called " secondary symptoms." A rigor, followed by nervous and vascular excitement, will seize on the patient, and on examination of the patient a few days after this occurrence, he will be supposed to be laboring under *measles*, in consequence of the cutaneous eruption which presents itself. After a further interval, the spots will be observed to be covered with copper-colored scales.

Explanation.

Although the ulcer is healed, yet the poison still remains in the part, and the organic glands in the site of the hardness are *impregnated* with the poison. The blood having to pass *through* the glands, after a certain time poisons the organic glands all over the body. The venous blood is carried to the right side of the heart: it receives its oxygen in the lungs; it is subsequently conveyed to the heart, and by the arteries all over the body. On the union of the oxygen with the glands, the syphilitic virus contained in the blood is *communicated* to the organic glands. When the glands are completely poisoned, they manifest the trouble that embarrasses them, by the rigor; they give evidence of their determination to resist the attack by the vascular and nervous excitement which supervenes on the rigor; and lastly, attempt to expel the poison from the system through the *agency* of the organic glands, which desirable end they *fail to accomplish*, in consequence of the poison being of a *specific kind*. In this respect *syphilis* is *totally* different from *small-pox* or *typhus fever*, either of which diseases will be *thrown off* by the powers of the organic nervous system, provided the patient survives a certain number of days.

Poisoning of the Organic Nervous System by Poison applied to a certain Part through the Blood.

" *Hydrophobia.* "

When a person is bitten by a rabid dog, the wound may heal, and remain so for weeks or months, and the part attract no attention until symptoms of hydrophobia set in, when the site of the original wound will be found inflamed. The poison having remained dormant for a certain period, at length is thrown into action; the organic glands of the part bitten attempt to throw off the poison by secretion, but the arterial blood, having to circulate through these glands, becomes impregnated with the poison; the venous blood, derived from the glands, must necessarily contain the poison, which, on being conveyed to the right side of the heart, and receiving oxygen in the lungs, is conveyed to the left side of the heart, and by the arteries all over the body. The organic glands are contaminated by the virus, on the union of the oxygen with the venous blood. The poison shows a particular predilection for the spheno-palatine ganglion, and selects it to carry into effect its fatal and dreadful operations, (as more fully explained in another place.) Why this should be the case, may be surmised when it is recollected that the saliva containing the poison in the dog is secreted under the morbid action of the spheno-palatine ganglion.

The conclusion to be arrived at is this—that the poison produced by the morbid condition of the *spheno-palatine ganglion in a dog* is capable of producing a similar condition of the *spheno-palatine ganglion* in *man*, by the inoculation of the poison.

Certain Medicines produce certain Effects on the Organic Nervous System, whether applied Externally or administered Internally.

" *Belladonna.*"

When belladonna is applied round the eyelids, the pupil is dilated. The iris is supplied with organic nerves exclusively; hence it is clear the belladonna acts on the organic nerves of the capillary arteries, and the glands in which these nerves terminate. Belladonna applied over the stomach, or round the nipple, or applied to the os uteri, or introduced into the rectum, will produce dilatation of the pupil, by its action on the organic nerves of these parts. Belladonna, when taken internally, acts on the organic nerves of the stomach; dilatation of the iris follows. It is evident, from what has been just stated, that belladonna acts on the organic nerves wherever applied, and simultaneously on the whole organic nervous system.

"*Opium.*"

When opium is taken into the stomach, contraction of the pupil takes place; when opium is applied to a part denuded of the cuticle by a blister, contraction of the pupil is the result; when opium is introduced into the rectum, contraction of the pupil is the result; when opium is injected in a state of solution under the skin, contraction of the pupil follows. The opium has the same influence over the organic nervous system wherever applied, and must act on the whole organic nervous system at the same time.

" *Strychnine.*"

When strychnine is taken into the stomach, tetanic spasm of the muscles is the result; when the cuticle is removed by the application of a blister, and strychnine applied to the abraded surface, tetanic spasm is the result; when strychnine is introduced into the rectum, tetanic spasm is the result. Strychnine is followed by the same effects on the organic nervous system wherever applied, and embraces the whole organic nervous system by its action.

" *Tobacco*."

When tobacco is taken into the stomach, it induces intense nausea, as well as great prostration and relaxation of the muscular system. When given as an enema, it produces the same effects. When applied to the head, it produces similar effects. The tobacco acts on the organic nervous system wherever applied, and on the whole organic nervous system at once.

" *Arsenic*."

When arsenic is taken into the stomach, it produces intense burning, extreme thirst, constriction in the throat, with vomiting and purging. When arsenic is applied to remove a cancer, or as a local application in cutaneous diseases, or introduced into the rectum, it is followed by the same symptoms or effects on the organic nervous system. It extends its action to the whole organic nervous system at the same time.

"*Acetate of Lead*."

A solution of GOULARD'S extract dropped on an inflamed conjunctiva will cause contraction of the organic nerves surrounding the capillary arteries.

Lead applied in the form of white paint to a scalded surface will cause contraction of the organic nerves surrounding the capillary arteries.

Lead administered internally in uterine hæmorrhage, or in hæmoptysis, or hæmorrhage from the bowels, causes contraction of the organic nervous system wherever distributed.

Explanation.

The lead acts as an irritant on the organic nerves surrounding the capillary arteries, causing *contraction* of the vessels, closing their mouths, preventing the ingress of blood into them; thus arresting the circulation of the blood in the capillaries, and consequently not only arresting hæmorrhage, but preventing the occurrence of the sequelæ consequent on an increased action of the organic nerves surrounding the capillary arteries, or what is commonly called inflammation.

Proofs that Lead acts as an Irritant on the Organic Nervous System.

The " dropped hand " that painters are afflicted with affords a proof that lead acts as an irritant on the organic nervous system, and causes rigid contraction of the muscles. A matter any one can satisfy himself of who attempts to extend the thumb and fingers of the " dropped hand," when he will find that opposition from the pronators and flexors impede his efforts to extend the fingers. In this case the supinators and extensors are in a normal state, whilst the flexors and pronators are in a state of spasm.

Medicines or Agents applied at distant Parts of the Body followed by the same Effects as if introduced into the Stomach, in consequence of their being conveyed into the Blood by the Capillary Veins and Lymphatics.

"Mercury."

When a person takes a certain quantity of calomel in the course of the day, after some time his gums will get tender, and the secretions from the salivary glands will be largely increased.

When a certain quantity of mercurial ointment is rubbed in the groin or to the calf of the leg for some days, a spongy state of the gums and salivary secretion will set in.

When a cinnabar is applied in the form of fumigation to the nose, after the lapse of a certain interval a spongy state of the gums and salivary secretion will be the result.

When a person is exposed to the vapors of mercury in mines or factories, a spongy state of the gums and salivation will be produced.

When a man is put under the influence of mercury in a badly-ventilated ward, he is liable to be attacked with erythema.

Explanation.

When a person is daily taking calomel, the mercury is brought in contact with the mesenteric glands. The glands become impregnated with the mercury, and impart it to the venous blood; the mercury is also taken up by the lacteals, lymphatics, and carried by the throacic duct into the venous circulation, which conveys it into the lungs, where the blood is oxygenized. The blood is conveyed next to the heart, from whence it is distributed to all parts of the body by the arteries, and to the glands, on the union of the oxygen with the latter; the organic glands of the salivary glands being more susceptible to the action of the mercury, become irritated, and make an effort to throw off the poison by the secretion of saliva.

When mercurial ointment is applied to the groin or calf of the leg, after some time the organic glands become imbued with the mercury; the blood becomes impregnated with the mercury, and is carried by the veins and lymphatics to the right side of the heart, from thence to the lungs; it is next conveyed to the left side of the heart, and from thence by the arteries all over the body, and to the glands, on the union of the oxygen with the former; the salivary glands give evidence of irritation, by their effort to remove the poison by the process of secretion.

When fumigation is had recourse to, the organic glands, in the Schneiderian membrane, become imbued with the mercury; the blood next becomes imbued with the mercury, and the salivary glands give evidence of irritation, in the manner already described.

When a man is exposed to the vapors of mercury in factories or mines, the vapor, being immaterial, is mixed with the air, passes into the blood with the oxygen, and is given off to the organic glands, 'on the union of the oxygen with them. After a certain interval the organic glands of the salivary glands endeavor to remove the poison in the manner already stated; namely, by the secretion of saliva.

Mercurial *erythema* is now a rare occurrence; formerly, when mercury was administered indiscriminately, it was of very frequent occurrence, particularly in what were called the *"foul wards"* of an hospital, where patients suffering under syphilis were using mercury in every shape and form, and where the atmosphere was poisoned by human effluvia and mercurial vapors. The mercurial vapor passes with the oxygen into the blood, poisons the glands, precisely in the manner as described in small-pox, typhus fever, and intermittent fever. The papillæ or vesicles which characterize the eruption indicate the efforts of the organic nervous system to throw off the poison by the process of secretion.

"Arsenic."

When arsenic is taken into the stomach, it causes intense burning, extreme thirst, constriction in the throat, vomiting, followed by purging, and great prostration of the organic nervous system.

When arsenic is applied to a cancer of the lip or elsewhere for too long a period, or when too great a quantity is applied at once, symptoms of poisoning by arsenic will present themselves.

When a lotion containing arsenic is applied for too long a period, with a view of curing a cutaneous eruption, symptoms of poisoning by arsenic may be anticipated.

When a man is employed in a paint factory where arsenic forms one of the chief ingredients in the paint, he may have symptoms of poisoning by arsenic, besides being afflicted with the loss of his nails, and having irritable ulcers on his fingers.

Explanation.

When arsenic is taken into the stomach, it acts as a direct irritant on the organic nerves of the stomach; the arsenic is found imbedded in the mucous membrane of the stomach, and consequently is in contact with the organic nerves. The arsenic acts as an irritant on the glands, which irritation is propagated to the nerves surrounding the capillary arteries, and to the muscular fibres of the stomach, through their connection with the capillary arteries. The stomach, therefore, contracts in order to throw off the poison—vomiting is the result. The organic glands of the stomach make a further effort to get rid of the poison by the process of secretion, and are assisted in their attempt by the mesenteric glands of the intestines, which at first secrete mucus, afterwards serum, and lastly blood, which is passed off by stool.

When the arsenic is employed in the form of ointment or lotion, the glands become impregnated with the arsenic; the venous blood of the part in which the arsenic is located becomes impregnated with the poison. The blood is carried to the right side of the heart, then to the lungs, where it is oxygenized; it is next conveyed to the left side of the heart, and from the latter by the arteries all over the body, and imparted to the organic glands, on their union with the oxygen. The poison shows a predilection for the organic glands of the stomach, and the symptoms of poisoning by arsenic are precisely the same as when it is taken into the stomach.

When a person is exposed to the vapors of arsenic, as in a paint factory, the vapor passes with the oxygen into the blood, and is given off to the organic glands all over the body, on the union of the oxygen with the former. The condition of the nails and fingers demonstrates this to be the case. The organic glands of the stomach being more susceptible of the poison than the organic glands in other organs, give evidence of the symptoms produced by the presence of arsenic.

"Tartar-Emetic."

When tartar-emetic is taken into the stomach, it irritates, and subsequently nauseates, the organic glands of the stomach.

When tartar-emetic is injected into the veins, it produces vomiting and nausea.

When tartar-emetic ointment is applied to the skin, it acts on the organic glands in the same manner as syphilitic poison, and is followed by a pustule.

Explanation.

When tartar-emetic comes in contact with the organic glands of the stomach, it irritates them; the irritation extends to the organic nerves surrounding the capillary arteries, and to the muscular fibres of the stomach, through the connection of the capillary arteries with the nerves; contraction of the stomach follows, and vomiting is the result. The gastric ganglion being cognizant of the nauseating tendency, debilitating action, as well as fatal results to be apprehended from the presence of the poison, endeavors to guard against its operation, by expelling it from the stomach by the process of vomiting.

When tartar-emetic is injected into the veins, it is carried to the right side of the heart, from thence to the lungs; it is next conveyed to the left side of the heart, and by the arteries all over the body, and is given off to the organic glands, on their union with the oxygen; on coming in contact with the gastric ganglion, it causes vomiting, in the same way as when it is administered by the mouth.

When tartar-emetic ointment is applied to the surface, it irritates the organic glands; the irritation extends to the organic nerves surrounding the capillary arteries; dilatation of the arteries follows; a greater quantity of arterial blood enters the capillaries. The glands now commence to throw off the poison by the secretion of lymph, serum, and pus; hence the pustule which follows.

"Brandy."

The effects of brandy are almost too well known to require description. It stimulates the organic glands distributed in every quarter. The heart pulsates stronger; respiration is more vigorous; animal heat is increased; the countenance is flushed, and the *eyes sparkle* with increased animation.

When brandy is thrown into the eyes, vessels containing red blood appear where none seemingly existed; the eyes, in truth, become what is popularly called "blood-shot."

Explanation.

The brandy acts as an excitant on the organic glands all over the body; increases the strength of the organic nerves wherever distributed; causes dilatation of the capillary arteries; allows more blood to enter them; supplies a larger quantity of oxygen to unite with the organic glands; invigorates the capillary circulation; heats the surface, by causing a greater consumption of oxygen by the organic glands.

The Action of certain Agents introduced into the Venous Blood from the Intestinal Tube, by the Lacteals and Thoracic Duct, on the Organic Glands of certain Organs.

When a person takes a drink of water, it rapidly passes into the venous blood, through the agency of the lacteals, lymphatics, and thoracic duct; the blood is carried to the right side of the heart, from thence to the lungs, and, on being oxygenized, is next conveyed to the left side of the heart, and from the heart by the arteries all over the body. The blood, on passing through the organic glands of the kidneys, has the water, in the shape of urine, secreted from the blood by the renal organic glands, on the same principle that the organic glands of the surface secrete the cutaneous perspiration. It will be recollected that the renal glands are formed by the plexus of nerves which surround the renal arteries derived from the renal ganglions, under whose presidency the urine is secreted.

"Gin."

When a man drinks half a pint of cold water with a glass of gin mixed with it, the power of the organic glands of the kidneys will be increased, and urine will be passed more freely. The gin passes into the venous circulation, through the agency of the lacteals, lymphatics, and thoracic duct; the blood containing the water impregnated with gin is carried to the right side of the heart, from thence to the lungs, where it is oxygenized. The blood is next conveyed to the left side of the heart, and from the latter it is distributed by the arteries to all parts of the body, and brought in contact with the organic glands, on the union of the oxygen with the former. The renal ganglions are apprised of the gin being in the blood, through their connection with the vascular membrane which surrounds them, which is composed of organic glands, formed at the termination of the arteries. The stimulus of the gin on the nerve-tubules of the ganglions passes through the nerve-tubules of the nerves to the organic glands of the kidneys, and increases their power of secretion.

"*Glauber Salts.*"

When a man drinks half a pint of cold water, with two ounces of Glauber salts dissolved in it, he will have copious discharges from the bowels. The water containing the salts passes by the lacteals, lymphatics, and thoracic duct, into the venous circulation; the blood is carried to the right side of the heart, from thence to the lungs, to be oxygenized; it is next conveyed to the left side of the heart, and from thence all over the body by the arteries. The mesenteric ganglion is apprised of the presence of the salts, through its connection with the vascular membrane which surrounds it; its purgative action is communicated to the nerve-tubules of the ganglion and to the tubules of the nerves which go to form the mesenteric glands, and thus increases their powers of secretion, as witnessed in the loose discharges which pass from the bowels.

"*Tartar-Emetic.*"

Let a person drink half a pint of water, containing two or three grains of tartar-emetic dissolved in it, and after some time he will vomit the contents of the stomach, and afterwards vomit bile copiously. The water containing the tartar-emetic passes into the venous circulation, through the lacteals, lymphatics, and thoracic duct; the blood is conveyed to the right side of the heart, from thence to the lungs, to be oxygenized; next to the left side of the heart, and from thence it is distributed to all parts of the body by the arteries, and is brought in contact with the organic glands. The tartar-emetic, on being communicated to the gastric ganglion, through its connection with the vascular membrane, (the organic glands of the membrane communicate with the nerve-tubules of the ganglion,) thus apprising it of the presence of the tartar-emetic, the ganglion directs the organic glands of the stomach, through the nerve-tubules of the nerves, to cause contraction of the stomach, and throw off its contents. The tartar-emetic is also brought

in contact with the hepatic ganglion in the same way, which
directs the secretion of bile by the hepatic organic glands.

Bile is secreted by the hepatic organic glands in the liver.
In corroboration of the truth of the organic glands being thus
engaged in the secretion of bile, it is a fact worthy of notice,
that the *vena porta*, which contains the impure blood of the
intestines, and from which the bile is secreted, is accompanied
by a plexus of *organic* nerves, derived from the hepatic ganglion—a part of the plexus which accompanies the *hepatic* artery—inosculation of the *nerves* surrounding the hepatic artery
and vena porta takes place in the hepatic lobule in the same
way that inosculation of the nerves surrounding the hypogastric arteries of the fœtus, (which correspond, so far as the quality of the blood is concerned, with the vena porta,) inosculate
with the nerves surrounding the uterine arteries, (which correspond, as regards the quality of the blood, with the hepatic
arteries,) in the placental lobule.

The Modus Operandi of certain Agents on the Organic Nervous System.

"Brandy."

When a stimulant, such as brandy punch, passes by the lacteals, lymphatics, and thoracic duct into the venous circulation,
and is carried to the right side of the heart, and from thence
to the lungs, where the blood is oxygenized, and is next conveyed to the left side of the heart, and from thence to be distributed by the arteries all over the body, the stimulus of the
brandy is communicated to the organic glands, on the union
of the oxygen with them. The organic ganglions, such as the
superior central, the cardiac, and the pulmonary, give ample evidence of the stimulus being communicated to the nerve-tubules,
(of which the ganglions are partly composed,) through the connection of the organic glands in the vascular membrane (which
surrounds the ganglions) with the nerve-tubules. The stimulus
of the brandy is further communicated to the retina of nerves
surrounding all the arteries. It will be recollected that the

organic nerves send twigs into the coats of the arteries; that
the nerves become *incorporated* with the tissue of the arteries;
the brandy stimulates the organic nerves on the external, middle, and internal surface of the arteries: hence the bounding
pulse.

"*Tartar-Emetic.*"

When a person takes a quantity of tartar-emetic in solution,
it passes by the lacteals, lymphatics, and thoracic duct into the
venous circulation, is carried to the right side of the heart,
and from thence to the lungs, where the blood is oxygenized;
it is next conducted to the left side of the heart, from whence
it is conveyed by the arteries all over the body to the organic
glands; on the union of the oxygen with the organic glands,
the tartar-emetic is communicated to the latter, as well as to
the gastric ganglion. The tartar-emetic is communicated to
the nerve-tubules of the ganglion, through its connection with
the organic glands in the vascular membrane, and to the cardiac and pulmonary ganglions in the same way. The nauseating
and debilitating action is also communicated to the organic
nerves on the external, middle, and internal coats of the arteries, as well as to the muscular fibres to which the arteries are
distributed, in consequence of the connection of the nerves
with the arteries; hence the universal prostration.

"*Chloroform.*"

When chloroform is inhaled, the vapor passes into the blood
with the oxygen; the anæsthetic agent is thus communicated
to the organic glands, on the union of the oxygen with the organic glands. After some time, the organic glands, as well as
the organic ganglions, give evidence of its presence; the anæsthetic agent is communicated to the nerve-tubules, which
enter into the organization of the ganglions, by their connection with the organic glands in the vascular membrane which
surrounds the ganglions; the operations of the ganglions are
thus interfered with: the superior central ganglion is placed
in a quiescent state; the cerebral glands are similarly circumstanced, and cease giving off the volatile agent to stimulate the
nerve-fibres or tubules of the brain. The operations of the

8

mind cease, and sleep is the result. The cardiac and pulmonary ganglions give evidence of the action of the anæsthetic agent by the feeble action of the heart and stertorous respiration. The anæsthetic operation of the chloroform is communicated to the external, the middle, and internal coats of the artery by the arterial blood, which is impregnated with the vapor of the chloroform; its action is still further extended to the muscular fibres, in consequence of the connection of the capillary arteries with the muscular fibres, causing their relaxation, and ultimately arresting the action of the pulmonary glands—a causality followed by *death.*

Conclusions.

It is now sufficiently demonstrated that the organic nervous system can be stimulated to the highest degree; can be depressed to the lowest degree; and that the operations of the mind, or animal nervous system, can be suspended through the operation of certain agents on the organic nervous system.

Inflammation, Depression, and Excitation of the Organic Nervous System.

Boerhaave inculcated (*Aph.* 375, *et seq.*,) that inflammation was caused by an obstruction to the free circulation of the blood in the minute vessels; and this obstruction, he supposed, might be caused by heat, diarrhœa, too copious flow of urine and sweat, or whatever could dissipate the thinner parts of the blood, and produce a thickness or viscidity of that fluid, when the latter did not exist before the production of inflammation; he imagined that the larger globules of the blood passed into the small vessels, and thus plugged them up. This circumstance was termed an *error loci.*

(Cullen.) "A spasm of the extreme arteries supporting an increased action in the course of them, may, therefore, be considered as the proximate cause of inflammation, at least in all cases not arising from direct stimuli applied; and even in this case the stimuli may be supposed to produce a spasm of the extreme vessels."

(John Hunter.) "The act of inflammation is to be consid-

ered as an increased action of the vessels, which at first consists simply in an increase of distention beyond their natural
size. This increase seems to depend upon a diminution of the
muscular power of the vessels, at the same time that the elastic
power of the artery must be dilated in the same proportion."

(Dr. THOMPSON.) "There are two hypotheses which at present divide the opinions of pathologists respecting the state of
the capillary vessels affected with inflammation. According to
the first of these hypotheses, the inflamed vessels are in a state
of increased action; according to the second, they act with
less force than the trunks from which they are derived."

(Dr. PHILIPP.) "Therefore I cannot help thinking the nature of inflammation appears sufficiently evident: the motion of the blood is retarded in the capillaries, in consequence
of the debility induced in them; an unusual obstacle is thus op-
posed to its motion in the arteries preceding them in the course
of the circulation, which are thus excited to increased action."

(Dr. HASTINGS.) "Inflammation consists in a weakened
action of the capillaries, by which the equilibrium between the
larger and smaller vessels is destroyed, and the latter become
distended."

(Dr. GENDRON.) "Increased action of the capillaries is the
primary cause of inflammation."

It is evident that the distinguished men whose opinions I
have given confined themselves to examining the effects, without attempting to explain the cause of what they termed primary inflammation: "There is no effect without a cause."
This axiom is equally true of primary inflammation. The
cause of the primary inflammation is to be found in the state
of the atmosphere and idiosyncrasy of the patient who is seized
with inflammation; as, for instance, when a person is exposed
to damp, cold, and fatigue, as well as kept in a sedentary position for some hours during the night, and is attacked with inflammation of any organ, the cause of the inflammation must be attributed to the circumstances and the position in which the
patient was placed. The organic nervous system becomes exhausted by the fatigue it suffers from exposure to wet combined
with cold.

This damp air, as well as any obnoxious vapors combined

with it, passes into the lungs, for the purpose of giving oxygen
to the venous blood. The air, under such circumstances, cannot be considered pure; therefore, when the oxygen is given
off to the organic glands, the impurities contained in the atmosphere are communicated to the organic glands. After some
hours the organic glands become depressed under the depressing influences to which they have been subjected, and manifest
their indisposition by a shivering or rigor; in due time, however, the organic nervous system arouses into action, to throw
off the danger which imperils it, and selects an organ or region
of the body to commence operations in, with a view of secreting the offending matter from the body by the action of
the organic glands. Mr. HUNTER's declaration or opinion
that inflammation should be deemed a healthy process, is true.

The same law regulates the organic nervous system, with
respect to discharging the offending matter which gave origin
to inflammation of a certain organ, as guides it in throwing
off the poisonous matter of small-pox; the organic glands
being the secreting organs in both cases to throw off the obnoxious element.

The organic nervous system in the primary stage of inflammation is "depressed," as evidenced by the rigor, the feeble
action of the heart, pallid countenance, and debility.

The organic nervous system, in the second stage of inflammation, from being depressed, becomes excited or "irritated,"
as evidenced by the burning heat of surface, the flushed countenance, the strong action of the heart, and strong, hard pulse.

The organic nervous system, in the third stage of inflammation, becomes "resolved;" it assumes its function of throwing off the irritation, by the process of secretion, through the
operation of the organic glands, or restoring the part to its
normal state.

"Pleuritis."

When a person is attacked with pleuritis, he firstly gets a
chill, accompanied with the usual symptoms of depression;
secondly, he is agitated with great vascular and nervous excitement; and thirdly, the vascular and nervous excitement
subsides, leaving him in a prostrate and oppressed condition.

117

"*Depression.*"

The organic nervous system is " depressed " in the first stage of pleuritis.

"*Excitation or Irritation.*"

The organic nervous system is excited or " irritated " in the second stage of pleuritis.

"*Resolution, with or without Effusion.*"

The organic nervous system is " resolved " in the third stage of pleuritis, or has thrown off the offending matter by the process of secretion into the cavity of the pleura, or has resumed its normal state.

"*Abscess.*"

When an abscess is about being formed in any region of the body, there is, firstly, depression of the organic nerves; there is, secondly, excitation or irritation of the organic nervous system; and thirdly, there is resolution of the organic nervous system.

"*Depression.*"

In the stage of *Depression*, the organic pulmonary glands are unable to give off vital fluid or electricity to unite the oxygen of the air with the venous blood in sufficient quantity; the rigor or cold surface of the body, the pallid countenance, and feeble action of the heart and arteries are thus accounted for.

"*Excitation or Irritation.*"

In the stage of *Irritation*, the pulmonary glands give off a larger quantity of vital fluid or electricity than normal; consequently, a larger amount of oxygen passes into the venous blood; consequently, a greater provision is made for heating the surface of the body, by the combustion or union of the oxygen with the organic glands: the burning heat of the surface and flushed countenance are thus accounted for. The oxygen stimulates the organic nerves, or the internal, middle, and external coats of the arteries, as well as the nerve-tubules of the ganglions, through their connection with the organic glands in the vascular membrane, (which surrounds them.) Thus the excited state of the heart and arteries is clearly accounted for.

" Resolution, with Effusion."

In the stage of *Resolution*, the organic nervous system becomes tranquillized, and selects the best course to adopt under the circumstances in which it is placed—namely, to remove the immediate cause of the trouble which threatens its well-being. Accordingly, in one case the organic glands commence to secrete or remove the poison, by the process of perspiration, through the organic glands, on the surface of the body; in another case, by the process of secretion of urine by the renal glands of the kidneys; in a third case, by the process of secretion of the organic glands of the pleura, into the cavity of the pleura; in a fourth case, by the process of secretion of the organic glands of the peritoneum, into the cavity of the peritoneum; in a fifth case, by the process of secretion of the organic glands of the serous membrane of the pericardium, into the pericardium; in a sixth case, by the process of secretion of the organic glands in the puogenic membrane of an abscess, into the cavity formed by the puogenic membrane for the reception of the pus.

Certain Poisonous Agents are capable of seizing on the Organic Nervous System for a certain Number of Days, when they are then expelled by the Power of the Organic Nervous System.

" Typhus Fever."

When the poison of typhus fever is introduced with the oxygen of the air into the arterial blood, and is communicated to the organic glands on the union of the oxygen with them, a certain train of symptoms follow, which give rise to the term typhus fever, and which disappear after a variable number of days, varying from fifteen to twenty-one days, (generally speaking.) The disappearance of the fever is accompanied by perspiration, or secretion of urine, or sleep, or a gradual amelioration of the symptoms, with elimination of the poison from the organic nervous system by degrees.

"Small-Pox."

When the poison of small-pox is introduced with the oxygen of the air into the arterial blood, and is communicated to the organic glands by the union of the oxygen with the organic glands, it is followed by certain constitutional symptoms, and the appearance of a cutaneous eruption in the form of pimples, on the fourth or fifth day, which become vesicles about the eighth day, and pustules filled with pus on the thirteenth day, which degenerate into crusts or scabs on the eighteenth day, in consequence of the evaporation of the fluid parts of the pus. The poison is then discharged by the organic nervous system.

"Measles."

The poison of measles is capable of being thrown off by the organic nervous system in a certain number of days.

"Scarlatina."

The poison of scarlatina is capable of being thrown off by the organic nervous system within a certain number of days.

It is evident, from what has been stated in reference to typhus fever, small-pox, measles, and scarlatina, that an immaterial poison, mixed with the air, is communicated to the organic nervous system, on the union of the oxygen with the glands; and that in each case it is of a specific or distinct kind, and can be only recognized by its effects, as it does not admit of chemical analysis.

Certain Poisonous Agents, on seizing on the Organic Nervous System, cannot be expelled without the Aid of other Agents.

"Intermittent Fever."

Marsh miasmata, contained in or mixed with the air in malarious districts, are communicated to the organic glands, on the union of the oxygen of the air with the organic glands.

When the organic glands are fully impregnated with the poison, they become depressed, and manifest severe indisposition, by great prostration, coldness of surface, and gnashing of the teeth. Soon, however, they awaken into action, and become irritated, as evidenced by the burning surface, flushed countenance, and strong pulsation of the heart and arteries; and ultimately become resolved or determined to throw off the poison, by the process of secretion by the organic glands of the skin; hence the perspiration that ensues. The poison, or its effects, are thus for a time removed, but the organic nervous system is subjected to another attack of the same kind, every twenty-four, thirty-six, or seventy-two hours. The organic nervous glands being still imbued with the poison, require a powerful agent, in the shape of quinia, to overcome its action, and restore the organic nervous system to a healthy condition.

Quinia cures intermittent fever by communicating strength to the organic glands, as well as neutralizing the poison, by entering into combination with it. The quinia, on passing into the venous circulation through the lacteals, lymphatics, and thoracic duct, is carried to the right side of the heart, and from thence to the lungs, by the blood, where the latter receives its oxygen: it is next carried to the left side of the heart, with the arterial blood, and communicated to the organic glands, on the union of the oxygen with the latter. The *quinia* is brought in contact with the poison, and supplants it; thus restoring the organic nervous system to its pristine state.

Certain Poisons, on being communicated to the Organic Nervous System, continue to hold Possession of it until the Death of the Patient.

"Phthisis."

When a person sleeps with another laboring under phthisis, and breathes the same atmosphere for some months, he is almost certain to have the poison of phthisis communicated to him. The air the phthisical patient respires becomes impregnated with the poison discharged from the ulcerated surface or cavity in his diseased lung; the patient, therefore, when in a close room, begets an atmosphere contaminated with a poison produced by himself, and one, too, susceptible of being communicated to other persons who respire the air in the room where he resides. The poison passes into the air-cells of the lungs, and is brought in contact with the organic glands, on the union of the oxygen with them, where it remains in a state of incubation for some months, when it commences its operations in a manner familiar to all pathologists.*

Organic Nerves on the External Parts of the Body Poisoned by Immaterial Agents mixed with the Air.

"Erysipelas."

When a patient is located in a ward in hospital at a period when erysipelas is prevalent in consequence of an ulcer or wound, he will probably be attacked with the disease called erysipelas; the immaterial poison contained in the air is communicated to the organic nerves exposed in the ulcer or wound. The erysipelas is ushered in by a rigor, (depression of the organic nervous system,) followed by excitement, (excitement of the organic nervous system,) characterized by heat, pain, redness, and swelling. Resolution at length takes place, and the poison is thrown off by the organic nervous system.

"Hospital Gangrene."

When several persons are crowded together in a badly-ventilated hospital where there is a vast amount of human effluvia, where animal flesh is suffering decomposition, persons whose limbs have been recently amputated will have the organic nerves on the surface of the stump poisoned by the immaterial poison mixed with the air, and give rise to the formidable and loathsome disease called "hospital gangrene," which is enough to generate a pestilence in any hospital where it ex-

* I think I may safely assert, that very few men have had greater opportunities of testing the truth of the poisoning by phthisis than I have had, and continue to have; and I have not the shadow of a doubt about phthisis being communicated from one person to another.

ists as a disease, and which is well known to be so susceptible of propagation from one wounded individual to another, and where change of air is indispensable to restore the patient to convalescence.

"Puerperal Fever."

When a woman is recently delivered, the interior of the wound presents a raw surface or internal wound. An immaterial poison, which becomes mixed with the air at certain periods of the year, poisons the organic nerves on the raw surface of the wound; causes depression of the organic nervous system; is followed by a rigor, and afterwards excitement of the vascular and organic nervous systems, which is called puerperal fever.

The Organic Nervous System of a Sound Person Poisoned by a Secretion derived from a Person apparently Sound.

"Syphilis."

To demonstrate that the syphilitic poison may be communicated by the semen secreted under the influence of the organic nerves of the testicle, whilst imbued with the syphilitic virus, to a woman during coition, by the poisoning of the organic nerves of the vagina and cervix uteri, I will briefly allude to two cases, where the semen actually communicated syphilis to sound and healthy young women. And here I will add, by way of parenthesis, they were not "French ladies," and were as respectable as the cases reported by Professors PARKER and PORTER. In fact, they were persons on whom the slightest shadow of suspicion could not rest.

Mr. C. was under my care for *syphilitica psoriasis guttata*. I ordered him compound decoction of sarsaparilla, with the bichloride of mercury. He got well, with the exception of two small condylomata near the verge of the anus. He now said he should get married in a few days. I told him I did not think it safe to do so, and advised him to consult some other person. Accordingly, he took the opinion of one of the ablest and best surgeons in the metropolis, who gave him a written note, stating there was no apprehension to be entertained about his entering into the marriage contract. It is to be noticed there was not the slightest appearance of a sore on the penis. In about nine weeks after the marriage, the lady, with whom I was well acquainted, applied to me with an ulcerated throat and the copper-colored scaly eruption. Knowing how matters stood, I made no allusion to the genital organs, fearing I might arouse suspicion.

Mr. I., a young man, applied to me, having ulcers on his legs and other parts of the body. He told me he was ten weeks married; that he had the "bad disorder" before he was married, but that "the doctor" said he was cured, and might get married.

I examined his virile organ, and found it all right. In consequence of stating his fear that his wife got the disease from him, I requested to see her, and found her with an ulcer in the throat, and the copper-colored eruption. I inquired if she had any ulcers on the genitals, and on her replying she had not, I made no further examination.

Irritation of the Organic Nerves on the Internal Organs, followed by Irritation on the External Parts of the Body.

"Urticaria—Putrid Shell-Fish."

When a person eats certain kinds of shell-fish, or fish in a putrid condition, he is apt to be attacked with a cutaneous eruption, commonly called nettle-rash. The irritation of the organic glands of the stomach, caused by the fish, extends to the organic nerves on the surface of the body. Here it may be remarked, that cutaneous eruptions are almost invariably connected with a deranged condition of the organic nerves of the stomach, or what is called Dyspepsia.

"Iodide of Potassium—Balsam Copaiba—Belladonna."

When a person has taken iodide of potassium for some time, he is apt to be seized with a cutaneous eruption, which I have elsewhere particularly described. The iodide of potassium passes into the venous circulation, through the lymphatics, lacteals, and thoracic duct, and, after being conveyed to the right side of the heart, mixed with the blood, and next to the lungs, where the blood receives its oxygen, it is next conveyed to the left side of the heart, and from thence by the arteries all over the body to the organic glands, to which it is communicated, on the union of the oxygen with the glands; hence the action of the organic glands to remove the poison produces the eruption.

Balsam of copaiba, when taken as a medicine, also occasionally irritates the organic nervous glands, on being communicated to them, on the union of the oxygen of the blood with the glands, causing the cutaneous eruption.

When belladonna is administered in large doses, its operations are manifested by a scarlet efflorescence on the skin. The poison is taken up by the lymphatics, lacteals, and thoracic duct, and carried into the venous circulation; on being carried to the right side of the heart with the blood, and from thence to the lungs, where the blood receives its oxygen, it is conveyed to the left side of the heart, and from the latter by the arteries all over the body to the organic glands, to which it is communicated, on the union of the oxygen of the blood with the glands; the poison irritates the organic glands: hence the scarlet efflorescence which supervenes.*

Certain Medicinal Agents Poison the Organic Nervous System invariably at certain Localities.

"Arsenic—Mercury—Tartar-Emetic—Ergot of Rye—Lead—Spanish-Flies."

When arsenic is introduced into the venous blood, as happens when it is applied to an ulcer on some part of the body, it selects the organic glands of the stomach for carrying into effect its destructive agency as a poison. The explanation consists in recollecting that the organic glands of the ulcer

---

* It appears, from the statements of authors, that belladonna acts as a prophylactic, by subjecting the organic nervous system to the influence of a medicine capable of inducing a disease similar to scarlatina; or, rather, rendering the organic nervous system, by its action, insusceptible to the influence of the immaterial poison which gives rise to the disease known as scarlatina.

become saturated with the poison, and that the venous blood becomes impregnated with the poison, which is carried to the right side of the heart, and from thence to the lungs, where the blood receives its oxygen; and is next conveyed to the left side of the heart, and from thence by the arteries all over the body to the organic glands, and that it is given off to the organic glands of the stomach, on the union of the oxygen with the blood in the glands.

"*Mercury.*"

When mercury is introduced into the venous blood, by friction on the surface, or through the medium of the lacteals, thoracic duct, and lymphatics, on the oxygen of the arterial blood being given off to the organic glands, the mercury selects the organic glands of the salivary glands, gums, and tongue for carrying into operation its poisonous agency.

" *Tartar-Emetic.*"

When tartar-emetic is injected into the veins, it selects the organic glands of the stomach for carrying into effect its poisonous agency, on the union of the oxygen of the blood with them.

" *Ergot of Rye.*"

When the administration of ergot of rye is persevered in for some time, mortification of the toes is liable to take place. The ergot poisons the organic glands of the toes; destroys vitality in the glands; hence the oxygen of the blood cannot unite with them, and death of the parts so implicated is the consequence.

" *Lead—Painter's Colic.*"

When lead is introduced into the venous blood in too great a quantity from time to time, it ultimately acts as a poison on the mesenteric glands, inducing the disease called painter's colic. The poison is given off to the mesenteric glands, on the union of the oxygen of the blood with the glands. Spasm of the organic nerves surrounding the capillary arteries, spasm of the capillary arteries, as a consequence of spasm of the nerves,

and spasm of the muscular fibres of the intestines, as a consequence of their connection with the arteries, follow.

"Spanish-Flies."

When Spanish-flies are taken in excess, they act on the organic nerves of the kidneys; the irritant is propagated to the
organic glands of the kidneys, on the union of the oxygen
with the renal glands, inducing violent irritation, followed by
bloody, or suppression of, urine.

"Elaterium."

Elaterium selects the mesenteric glands for operation, and
causes such a discharge of serum as to interfere with the functions of the blood in carrying the oxygen from the lungs to the
organic glands.

"Opium."

Opium selects the cerebral glands; arrests their action, induces sleep, and ultimately destroys vitality, by its deadly
influence on the organic nervous system.

Loss of Vitality, or Death of the Organic Ner-vous System, caused by Want of Blood to convey the Oxygen to the Organic Glands.

"Hectic Fever—Diseased Joints."

When a person is suffering from hectic fever concomitant
on a diseased joint; when there are profuse night-sweats, al-
ternating with diarrhœa, together with a profuse discharge
from the diseased joint, as is exemplified in scrofulous disease
of the knee or hip joint, the quantity of blood gradually diminishes, until there is not enough left to convey a sufficient
quantity of oxygen to the organic nervous glands to hold life
in them.

"Phthisis—Pneumo-Thorax."

In the last stage of phthisis, when the patient is reduced to
a skeleton by the combined operation of night-sweats, diar-

rhœa, and expectoration of purulent matter, he will be suddenly subjected to a severe shock, caused by the entrance of air through an ulcerated opening into the pleura, causing collapse of the lung, and filling the place occupied by the lung with air. Under such circumstances, there is only one lung to furnish the oxygen. There is barely enough of blood left to convey as much oxygen as will hold life in existence. As the quantity of blood continues to diminish, the oxygen continues to diminish, until at length there is not enough to hold life in existence, and the patient dies for the want of oxygen.

"*Hæmorrhage.*"

The want of blood to convey the oxygen to combine with the organic nervous glands is indicated by the gasping for breath, which is caused by the violent action of the organic pulmonary glands to supply oxygen to the organic glands to maintain life; the cold surface demonstrates there is no oxygen to combine with the organic nervous glands; whilst the convulsions that ensue show the struggle that life makes before its departure from its abode in the organic nervous system.

"*Dropsy.*"

As dropsy is the effect either of diseased heart, lungs, liver, or kidneys, it is merely intended to show cause of death by ascites, where a patient has been subjected to paracentesis several times. In cases of this kind, all the serum of the blood is drained off, until there is not enough of blood left to convey the oxygen to the organic nervous glands; hence the patient eventually dies, for want of oxygen to combine with the organic glands. Previous to death closing the scene, the patient becomes emaciated to the lowest degree; verifying the truth of this explanation.

"*Rupture of an Aneurism.*"

When an aneurism is ruptured, the blood is directed out of its natural course; consequently, none is sent to the lungs, and the patient dies for want of blood to convey the oxygen to the organic nervous glands. The coldness of the surface, the pale countenance, and, in some instances, the speedy or instantaneous dissolution of the patient, are readily explained.

"Chronic Diarrhœa."

During the famine years, the poor people of Ireland were compelled to subsist on vegetables or garbage, as well as to drink copiously of cold water to allay thirst; hence in a short time they were attacked by diarrhœa, which closely resembled the action of mild aperients continually administered; in a short time they became weak, pale, and emaciated. In this condition they applied to the relieving officer, who either gave them out-door relief, or had them admitted into the workhouse. The diet in the workhouse for an able-bodied man consisted of sixteen ounces of Indian meal and one quart of buttermilk; thus giving eight ounces of meal for stirabout in the morning, with a pint of buttermilk, and the same quantity of meal for stirabout and a pint of milk for dinner. No supper was allowed. This diet, instead of arresting the diarrhœa, in many cases brought diarrhœa on. The poor people invariably applied to the medical officer for change of diet, namely, bread and milk, which improved their condition; in some cases, however, notwithstanding the change of diet, and the liberal administration of stimulants and proper nutriment, the patient continued to crave for more food and drink, yet became amazingly emaciated; the discharge from the bowels of a fluid resembling water in which cabbage had been boiled, continuing to harass the patient several times in the twenty-four hours. The patient would remain probably for some weeks in the condition described, when he would request admission into the Infirmary, being no longer able to sit up or walk about in his ward. The patient at this time would present a haggard countenance, sunken eyes, clammy skin; either no pulse at the wrist, or one scarcely perceptible; on examining his abdomen, the umbilicus would appear as if resting on the body of the lumbar vertebræ; in other words, the abdomen would seem completely empty of its contents, and the whole contour of the patient would remind the observer of a living skeleton. Here, it is to be remarked, the patient might survive for four or five days without the pulse being perceptible at the wrist, with the action of the heart exceedingly feeble, respiration scarcely perceptible, and the surface of the body cold; the patient at length dying without a struggle. The cause of death is now easily explained. The organic pulmonary glands daily losing the power to give off electricity to unite the oxygen of the air with the venous blood; the quantity of blood daily reducing, to convey the oxygen to the organic nervous glands, and eventually ceasing to convey any oxygen, accounts for the cause of death.

"*Asiatic Cholera.*"

When a person is attacked with Asiatic cholera, after a few copious serous discharges, the surface of the body will become deadly cold; respiration and circulation extremely feeble; the urine will become suppressed; (in consequence of the diminution of the blood, there is no provision made for the secretion of urine,) and the voice will be lost. Vitality in the organic nervous system is here prostrated to the lowest degree, and ultimately becomes extinguished, for want of blood to convey the oxygen to the organic glands.

Loss of Vitality, or Death of the Organic Nervous System, produced by Irritation of the Organic Nervous System.

"*Convulsions—Epilepsy.*"

Tickling the soles of the feet, irritation of the genital organs by masturbation, irritation of the gums by dentition, irritation of the mucous membrane of the intestines by worms or irritating food, will produce convulsions. Here the convulsions are the result of excitation of the organic nerves of the parts specified. No person can attribute the convulsions produced by irritation of the bowels to the action of the animal or cerebrospinal nervous system, inasmuch as the intestines receive no nerves from this source.*

---

* Irritation, as in the case of the temporal artery when the wound is twitted, will produce alternate contraction and dilatation of the organic nerves surrounding the arteries, if persistent irritation be kept up, as exemplified when a child is cutting a tooth; contraction and relaxation of the nerves surrounding the capillary arteries on which the tooth is pressing must be the conse-

The mode in which death is caused by tickling the feet will be understood by recollecting that continued laughter is kept up; that the organic pulmonary glands become so exhausted as to be unable to give off electricity to unite the oxygen with the venous blood.

" *Delirium Tremens.*"

When a man addicted to habits of intemperance for several years, and who has had three or four attacks of delirium tremens, gets the disease for a fifth time, his animal and organic nervous systems will be found greatly deranged. The derangement of the former is indicated by spectral illusions and erroneous ideas. The derangement of the latter is made manifest by the general tremor of the muscles, the soft, weak pulse, and the fluttering of the heart. A patient thus circumstanced may prolong life for a few days, when death will close his career either by convulsions or sudden dissolution. I do not include coma, as I think when a patient dies in this condition, it is very often from overdoses of narcotics. The pulmonary organic glands fail by degrees to give off enough of electricity to unite the oxygen with the venous blood to hold life in a vigorous condition; hence the tremor of the muscles; hence the weak action of the heart and arteries; hence it is that if venesection is practiced the doom of the patient is sealed, inasmuch as the removal of the blood removes the medium of conveying the oxygen to the organic glands, as well as depresses the organic nervous system by its sedative action.

Exhaustion of the organic nervous system, consequent on irritation of the organic nervous system, accounts for death.

---

quence. When the irritation becomes very great, it is propagated to the entire organic nervous system, that is to say, to the organic nerves surrounding the arteries all over the body, as well as to the muscles to which the arteries are distributed; hence the alternate relaxation and contraction of the muscles which follow the disturbance is propagated to the brain through the organic nerves surrounding the arteries at the base of the brain, and distributed to the peripheral surface of the brain; hence the operations of the mind cease, and the person becomes insensible to external objects.

Loss of Vitality, or Death of the Organic Nervous System, caused by Obstruction to the Entrance of Oxygen into the Lungs.

" Bronchitis."

Death by bronchitis, where the bronchial tubes are filled with mucus, is produced by the obstruction given by the mucus to the entrance of the air into the lungs, whereby the operation of the pulmonary organic glands is interfered with, and the consequent union of the oxygen with the blood interrupted; hence death is caused by the want of oxygen to combine with the organic nervous glands.

" Tetanus."

When a person gets a punctured or lacerated wound in the palm of the hand or sole of the foot, he is liable to be attacked with spasm of the muscles of the face, neck, and trunk after some days. After the lapse of three or four days from the date of the attack, the patient generally dies. Here spasm of the arytenoid muscle closes the glottis, so that no air can enter the lungs; no oxygen can consequently be given to the venous blood. Death therefore takes place from the want of oxygen to combine with the organic nervous glands.

" Strychnine."

When a large quantity of strychnine is taken into the stomach, or applied to a blistered surface, or injected into the rectum, spasm of the muscles of the neck, trunk, and extremities will be the result, accompanied with great agony. Here the strychnine acts as an irritant on the organic nervous system, and death closes the scene by closure of the glottis, produced by the spasmodic action of the arytenoid muscles, which close the glottis, thus preventing the entrance of air into the lungs. The patient, therefore, dies for the want of oxygen to combine with the organic nervous glands.

" Spasm of the Glottis."

In this formidable disease, the glottis is closed by the spasmodic action of the arytenoid muscles. No air can enter the lungs; therefore death is caused by the want of oxygen to combine with the organic glands.

Loss of Vitality, or Death of the Organic Nervous System, consequent on an Injury of the Organic Nervous System.

A strong Proof that Life is located in the Organic Nervous System.

"Fracture of Cervical Vertebræ."

When a man, in consequence of a fall from a height, gets a fracture of the third cervical vertebra, he will be found unable to move any part of himself, with the exception of his head. His will cannot influence the motion of his extremities, as the spinal cord, the internuncio, suffers by the pressure of the fractured bone. Generally, in from two to four days, the patient dies suddenly; perhaps whilst eating or drinking, or at a time when least expected.

Explanation.

It will be recollected the cardiac nerves which arise from the cervical ganglions go to form the cardiac ganglions; any violence done these nerves is communicated to the cardiac ganglions; hence the sudden cessation of the heart's action, as well as the sudden cessation of the functions of the pulmonary organic glands, derived from the pulmonary ganglion. Death is thus caused by the want of oxygen to combine with the organic nervous glands.

Loss of Vitality in the Organic Nervous System, consequent on Injury of Organic Nerves at a distant Part of the Body.

"Compound Fracture—Mortification."

When a man gets a compound fracture of the leg, caused by the wheel of a vehicle passing over it, the organic nerves surrounding the external, middle, and internal coats of the arteries are contused; their vital functions immediately cease; the union of the oxygen with the organic glands is interrupted;

death of the parts below the injury, or mortification, follows
as the result.

Explanation.

The organic nervous system suffers from the deadly shock,
as indicated by the Hippocratic countenance, the cold surface,
the absent or intermittent pulse, the feeble respiration, the
hiccup, and sudden dissolution of the patient. The pulmonary
organic glands fail to give off electricity or vital fluid to unite
the oxygen of the air with the venous blood, and death closes
the scene.

Certain Agents destroy Life by their Sedative Action on the Organic Nervous System.

"Hydrocyanic Acid."

When a drop of concentrated hydrocyanic acid is placed
on the tongue, it acts as a direct sedative on the organic nervous system, and expels vitality in every part of the organic
nervous system, resulting in instant death.

"Veratrum Viride."

When veratrum viride is taken in an overdose, it acts as a
direct sedative on the organic nervous system, as indicated by
the cold surface, the intermittent pulse, the feeble respiration,
the prostration of the vital powers, and death of the organic
nervous system. The poison destroys life by its sedative action on the organic nervous system.

"Tobacco—Depression and Death of the Organic Nervous System."

That tobacco given as an enema or taken into the stomach,
or applied extensively to the surface of the body, will produce
intolerable nausea and vomiting, accompanied by extreme
prostration, pale countenance, feeble pulsation of the heart,
complete relaxation of the muscles, and sudden death, is well
known.

Explanation.

The tobacco poisons or acts as a depressing agent on the
organic nerves of the stomach, on the organic nerves of the

heart, on the organic pulmonary glands; depressing the latter,
so as to be unable to give off electricity to combine the oxygen of the air with the venous blood. The coldness of the surface is caused by the want of oxygen to unite with the organic
glands; the sudden dissolution of the patient is caused by the
want of oxygen to combine with the organic glands. All
these phenomena are attributable to the nauseating and depressing influence of the tobacco on the whole organic nervous
system.

"Cold—Death of Organic Nervous System."

When a person is exposed to intense cold in a sedentary
position, the temperature of the body and extremities quickly
falls below the natural standard; the cold acts as a sedative
on the capillary organic nerves, causing contraction of these
vessels, and consequently preventing the entrance of the blood
containing oxygen to unite with the organic glands. The cold
air, in its passage into the lungs, after some time, and by degrees, paralyzes the pulmonary glands, so as to render them
unable to give off enough of electricity to cause the oxygen
of the air to combine with the venous blood; hence it is that
the person dies in an imperceptible manner, the union between
the oxygen and the venous blood in the lungs having ceased.

" Heat—Death of the Organic Nervous System."

In certain electrical conditions of the atmosphere, persons
who are exposed to the heat of the sun's rays, or persons employed in lofts under the surface of the earth, suddenly fall
dead, or have animation suspended for some time. Vitality
becomes so exhausted in the organic nervous system that it
suddenly fails, and the pulmonary organic glands being unable
to give off electricity or vital fluid to unite the oxygen of the
air with the venous blood, the patient dies for the want of oxygen, consequent on exhaustion of the organic nervous system.

" Shock—Death of Organic Nervous System."
"Scald."

When a person is extensively scalded, he will be soon found
chilly, and afterwards be attacked with a rigor; the shock

given to the organic nerves of the skin is communicated to the whole organic nervous system, as evidenced by the weak action of the heart and feeble respiration. The pulmonary organic nervous glands fail to give off electricity in sufficient quantity to unite the oxygen of the air with the venous blood; and if the shock is very severe, the pulmonary organic nervous glands fail *in toto* to give off oxygen, and the person dies for the want of oxygen, consequent on the shock given to the organic nervous system.

"*Intoxication—Death of Organic Nervous System.*"

When a person drinks largely of intoxicating liquors, the whole organic nervous system is thrown into a state of unusual excitement, as well as the animal nervous system, as indicated by the gesticulation and by the tottering gait of the individual when he attempts to walk. The will commands the lower extremities to walk, but stumbling is the result, in consequence of the central superior ganglion being unable to guide or regulate the action of the prevertebral ganglions, on whose action the regular contraction of the muscles depends to suit the requirements of the mind. If the stimulus continues to be imbibed to excess, the organic pulmonary glands become exhausted, and by degrees lose the power of giving off electricity to unite the oxygen of the air with the venous blood; all power soon ceases, and the person dies for the want of oxygen to unite with the organic glands to hold life in existence.

"*Brandy.*"

Brandy in moderate doses acts as a stimulant on the organic nervous system, but in excessive quantities as a deadly sedative.

"*Uræmia—Cause of Death—Poisoning of the Organic Nervous System.*"

When a patient has been suffering for some time under Bright's disease of the kidneys, whose urine is largely charged with albumen, as well as free from urea, he is liable at any time to be attacked with convulsions, to be followed by coma and death. The urea contained in the blood is given off to the organic glands, on the union of the oxygen with the glands. The poison irritates the organic glands; the irritation is propagated to the arteries, through their connection with the organic nerves which surround them, on the internal, external, and middle coats; the irritation is propagated to the muscles through the arteries with which the muscles are connected, as well as to the muscular fibres; hence the alternate contraction and relaxation of the muscles which supervene. When coma supervenes, it is caused by the suspension of the functions of the cerebral glands. No volatile agent being secreted, sleep must ensue, in the manner already explained. The urea acts as a poison on the organic nervous system, and ultimately destroys life.

"Jaundice—Poisoning of the Organic Nervous System by the Bile."

When the bile is not secreted by the hepatic glands of the liver, or when it is absorbed and gets into the blood, after being secreted, the skin becomes intensely yellow. Sometimes the patient is attacked with violent delirium, convulsions, and dies in a state of coma. The bile is communicated to the organic glands, on the union of the oxygen with the glands. The bile acts as an irritant in the first instance, inducing spasm of the organic glands; which condition of the organic glands is propagated to the organic nerves surrounding the arteries on their external, middle, and internal coats: alternate relaxation and contraction of the coats of the arteries is the result. As the arteries are distributed to the muscular fibres, alternate contraction and relaxation of the muscular fibres must ensue.

In this manner can be explained the alternate contraction and relaxation of the muscles, which give rise to the name of "convulsions." The delirium is caused by the volatile agent being generated by the cerebral glands, whilst the latter are suffering from the irritation caused by the poison. The coma which sets in is caused by the suspension of the action of the cerebral glands, and consequently the arrest of the secretion of the volatile agent, on whose action the operations of the mind depend. Death is caused by loss of vitality in the organic nervous system, induced by the poison contained in the bile.*

The Object of Therapeutic Agents is the Restoration of the Organic Nervous System to a Healthy State.

"Typhus Fever."

In the treatment of typhus fever, *gravior* or *mitior*, and in fact all the eruptive fevers, it is to be recollected, if the patient can be kept alive for a given number of days, that his chances of recovery are exceedingly good. No matter how violent the symptoms are which characterize the fever, it should be always kept in remembrance that the organic nervous system is *suffering* from a *poison*, and will require a certain number of days to throw it off; and that, instead of depressing the organic nervous system by the abstraction of blood, by leeches, cupping, venesection, or purging, the strength of the patient should be

---

* The late Sir Henry Marsh, in the third volume of the Dublin Hospital Reports, gives the particulars of a case which came under the notice of the late distinguished Mr. Colles. "A young gentleman, having a chancre on the glans penis, went to his house to consult him. He directed him alterative doses of calomel, which were persevered in for four or five weeks. The mercury seemed to agree well. No untoward symptom appeared, and the ulcer was completely healed. About three weeks afterwards the young man was observed to be deeply jaundiced; and having continued in this state two or three days, he was suddenly seized with delirium, followed by repeated convulsions. These symptoms having continued for a few days, Mr. Colles was sent for, and found his patient dying. The symptoms indicated, evidently, a most violent affection of the brain. *Every viscus* in the body was most accurately *examined*, and not a trace of disease *could be discovered*. The external and internal parts were much tinged with bile."

Morgagni mentions the case of a young priest, "who, soon after perturbation of mind, was seized with jaundice; pain in the epigastrium; vomiting; the stools colorless. After a few days he was restless, stupid, and forgetful; then delirious and convulsed. He gnawed everything with his teeth, struggled violently, and vomited dark matter. The blood rushed impetuously from an opening made in a vein! the serum gave the lunar rays a yellow tinge; the convulsions ceased; he lay motionless and comatose, and died on the fifth day."

These cases illustrate the truth of the doctrine, with respect to the mode in which the *bile poisons* the *organic nervous* system, which I have put forward.

carefully guarded, inasmuch as it will be required towards the
termination of the disease. Interference in the treatment of
the diseases specified requires great judgment on the part of
the physician to know when interference is required, or when
it is injurious. In some very bad cases of typhus fever, the
prompt and energetic treatment adopted by the physician is
crowned with the most glorious results. The marvelous effects
which follow the administration of brandy punch, beef-tea, combined with muriate of soda, are calculated to inspire confidence
in the mind of the physician. The resuscitation of the organic
nervous system, as indicated by the increased action of the
heart and arteries, the tranquil respiration, the returning animation of the countenance, the recovery of the mental faculties, the subsidence of the tympanitic abdomen, the discharge
of flatus, and the condition of the tongue, all demonstrate the
utility and advantage of the treatment adopted.

"*Acute Inflammation of Important Organs.*"

In all the cases of acute inflammation of important organs
occurring in persons of vigorous constitutions, where the pulse
points out that the organic nervous system is suffering from irritation, venesection should be carried into effect at once, and
repeated according to circumstances; the blood-letting to be
followed with an anodyne draught, which is to be repeated if
necessary.*

"*Collapse.*"

When a man receives a great shock by a fall from a height,
whose respiration and circulation are extremely feeble, whose
surface grows rapidly cold, and whose countenance is pale and
depressed, he should be placed on his back; his surface should

* When an inflamed organ is about throwing off the irritation by the effusion
of lymph. serum, or pus, blood-letting should not be practiced, nor should purgatives be administered. The patient's strength must be preserved by light,
nutritious regimen, and occasionally stimulants. Mercury is the sheet-anchor
for removing the effects of inflammation at this stage, and must be persevered
in until its physiological effects are produced, when all the signs of inflammation
will disappear, and the organic nervous system be relieved from oppression and
danger.

be smartly rubbed, his face and nostrils washed with brandy; and, in addition, he should get some brandy, diluted with water, to drink, and afterwards some beef soup. The object of the treatment is to restore the organic nervous system to its former strength, and enable the cerebral, the pulmonary, the cardiac ganglions, as well as the organic glands all over the body, to discharge their functions, and again establish mental intelligence, increased activity of the circulation and respiration, accompanied by increase of the temperature of the surface of the body.*

"*Paraplegia—Strychnine.*"

When a person labors under paraplegia, brought on by excessive irritation of the organic nerves of the genital organs, shower-baths, frictions, and strychnine are had recourse to, in order to restore the muscles to their former functions. After the treatment has been persevered in for some time, the patient will complain of spasms of the muscles attacking him at intervals, with more or less improvement of muscular strength on the cessation of the spasms.

Explanation.

The strychnine passes into the venous circulation through the agency of the lymphatics, lacteals, and thoracic duct; it is carried to the right side of the heart with the blood, and from thence to the lungs, where the blood receives its oxygen; it is next conveyed with the arterial blood to the left side of the heart, and from thence it is conveyed all over the body to the organic glands, to which it is communicated, on the union of the oxygen with the glands; it is also communicated to the organic nerves on the internal, middle, and external coats of the arteries, in its transition with the blood through the arteries.

---

* In former times it used to be the practice to bleed a man who had received a fall on a race-course, or who fell from the top of a house. Fortunately, however, for the sufferer, the operators could not accomplish their object, in consequence of the cessation of the circulation during the stage of collapse. Removing the blood, under such circumstances, would be removing the oxygen, on whose stimulus life is held in existence; a practice justly condemned by the celebrated SIR ASTLEY COOPER.

The strychnine irritates and causes spasm of the organic nervous system all over the body, if given in a large dose, but is moderate in its action when medicinally administered. The strychnine induces spasm of the muscles in the following manner:

The muscular fibres of the muscles are supplied with blood by the capillary arteries; the capillary arteries are surrounded on their external, middle, and internal coats with organic nerves. Contraction and spasm of the nerves must be followed by spasm and contraction of the capillary arteries, and spasm and contraction of the capillary arteries must be folowed by spasm and contraction of the muscular fibres.

Uterine Hæmorrhage.

"Ergot of Rye."

Relaxation of the muscular fibres of the uterus, after the expulsion of the placenta, is sometimes attended with profuse flooding. Obstetricians place reliance on the efficacy of ergot of rye in causing contraction of the muscular fibres of the uterus, and thus arresting the hæmorrhage, by the pressure of the muscular fibres of the uterus on the bleeding vessels.

Explanation.

When the ergot is administered, it is quickly conveyed into the venous circulation by the lymphatics, lacteals, and thoracic duct; it is next carried, with the blood, to the right side of the heart, and from thence, with the blood, to the lungs, where the latter is oxygenized; it is next carried, with the blood, to the left side of the heart, and from thence by the arteries to the organic glands all over the body, to which it is communicated, on the union of the oxygen with the glands. It causes contraction of the uterus in the following way:

The muscular fibres of the uterus are supplied with blood from the capillary arteries; the capillary arteries are surrounded on their external, middle, and internal surfaces with organic nerves. The ergot causes spasm of the nerves; spasm of the capillary arteries follows spasm of the nerves, and spasm of the muscular fibres follows spasm of the arteries.

Ileus, or Internal Strangulation of the Intestinal Tube.
" *Tobacco.*"

When a person is suddenly attacked with pain in the abdomen, accompanied by vomiting and obstinate constipation of the bowels, the probability is that he suffers from internal strangulation of the intestines. Under such circumstances, a tobacco enema is thrown up the rectum, with a view of removing the spasm and freeing the bowels of their contents, and sometimes with a fortunate result.

Explanation.

The tobacco is conveyed by the lymphatics into the venous circulation; it is next carried to the right side of the heart, with the venous blood; from thence it is conveyed to the lungs, where the blood receives its oxygen; it is next carried, with the blood, to the left side of the heart, and from the latter, by the arteries, all over the body, to the organic glands, to which its nauseating and deadly qualities are communicated. The nauseating and deadly qualities are also communicated to the organic nerves on the internal, middle, and external surface of the arteries, on the passage of the blood through the arteries. The organic nervous system is prostrated and nauseated to the lowest degree. The vital action is shaken in the organic nervous system to its very centre.

The muscular fibres are supplied with blood from the capillary arteries; the capillary arteries are surrounded by organic nerves, on their internal, middle, and external surface. The organic nerves are prostrated and nauseated; the capillary arteries must be similarly circumstanced, in consequence of their connection with the arteries, as well as the muscular fibres, to which the arteries are distributed.

Restoration of Asphyxiated Infants.
" *Cold Air.*"

In a medico-legal point of view, it is a very desirable matter to be able to explain why an infant, that an *honest midwife* or *indiscreet lady* has roughly thrown in a state of complete nudity into a sink, water-closet, or coal-hole, is subsequently found crying by an individual, who *charges* the midwife or other persons with a murderous intent to destroy the infant's life. It is quite true that an infant may be born to all appearance dead, and may continue in that state for a longer or shorter period, and eventually, under judicious management, be restored to life; it is also well known that often infants, after the usual treatment to restore animation was had recourse to, and apparently in vain, have been restored to life on being thrown in a cold room.

In a case such as the one above described, the infant is in a state of extreme exhaustion and vital prostration; on the body of the infant being exposed to the draught of cold air, all the organic nerves become constricted, firm, and strengthened; the arterial trunks and their capillaries become constricted, so that the blood is forced out of them; the muscular fibres of all the muscles, together with the muscular fibres of the heart, are contracted, in consequence of the connection of the organic nerves with the muscular fibres. It will be remembered, contraction of the capillary arteries cannot take place without causing contraction of the muscles to which the arteries are distributed, in consequence of the connection between the capillary arteries and organic nerves which surround them; contraction of the arteries with their capillaries, as well as contraction of the muscles, is followed by increasing the quantity of blood in the veins; the venous blood is sent by the contraction of the muscles to the right auricle of the heart; the auricle now contracts through the agency of the cardiac nerves from the right auricular cardiac ganglion; the blood passes into the right ventricle; the latter next contracts and sends the blood by the pulmonary artery and its branches to the pulmonary organic glands, which are formed by the retina of organic nerves surrounding the pulmonary arteries at their terminations, and which are derived from the pulmonary ganglion, which is placed in juxtaposition with the cardiac ganglion; as soon as the blood reaches the glands, the infant gives evidence of its having done so, and the pulmonary glands, true to their office, demand air; the child opens its mouth, makes an inspiration, the air rushes in; on the air coming in contact with the organic pulmonary glands, it stimulates the glands, which give off electricity, which causes the union of the oxygen of the air with the venous blood as it is circulating through the pulmonary organic glands; the blood, being now charged with oxygen, is conveyed to the left side of the heart, and from the heart all over the body, by the arteries, to the glands, to which they supply the oxygen required.

The immaterial agent known as life is now enkindled, and commences its operations all over the body, as fully announced by the crying of the child. Thus it will be perceived that the offspring of the poor and unfortunate are restored to life by exposure and privation, whilst those of the rich are doomed to certain death, engendered by the hot-bed of luxury which surrounds them, on being placed under similar circumstances as regards the asphyxiated condition described. An infant supposed to be dead, in the one case, is wrapped up in warm flannel, placed in a warm bed, in a warm room, to be looked at and gazed on as an untold loss, until it sleeps quietly in death; whilst in another case, the parties about the infant are only too happy that the little innocent has escaped the miseries of this wicked world, and summarily throw its body into such a place as will cause the infant to wake up into life.

"A Scald—Cold Water."

One of the quickest ways to produce vesication is to apply a sponge soaked in very hot water, for a second, to the surface. Heat, pain, redness, and swelling almost immediately present themselves. Vesication quickly follows. The application of cold water to the scalded surface, if at once had recourse to, and kept up for some hours, will allay pain and prevent vesication.

Explanation.

The hot water irritates the organic nerves surrounding the capillary arteries on their external surface, as well as the organic glands; dilatation of the arteries instantly takes place; a greater quantity of blood, with a larger quantity of oxygen, is thus permitted to flow into the arteries. The glands being furnished with a larger supply of oxygen and blood, at once commence to secrete the serum of the blood which passes through the pores of the skin and lodges beneath the cuticle, (the pores in the cuticle being closed by the heat of the application.) The cold water (or iced water, if convenient,) allays the irritation of the organic nerves and organic glands, causes contraction of the organic nerves surrounding the capillary arteries, prevents the admission of blood, and consequently of oxygen, into them; thus leaving the organic glands in a quiescent state, and thus preventing the secretion of serum.

" *Pleuro-Pneumonia—Venesection.*"

When a robust young man, after suffering fatigue, sleeps in a damp bed, he is liable to be attacked with shivering, followed by excitement of the nervous and vascular systems, as well as by a severe pain in the side, and difficulty of breathing. On stethoscopic examination, a slight rubbing sound will be heard at a point corresponding to the pain, as well as a very fine crepitus; on percussion, slight dullness at this stage of the disease will be perceptible. The patient is suffering from pleuro-pneumonia, and recourse should be had at once to bloodletting, and a sufficient quantity of blood abstracted to cause fainting or suspended animation, and thus give a decided shock to the organic nervous system. The patient should next get from thirty to forty drops of laudanum; in case the pain in his side or the difficulty of respiration continues, the lancet must be again had recourse to, as well as the anodyne draught; should effusion take place into the pleura, or hepatization of the lung ensue, the case must be treated on the principles laid down in another place.

Explanation.

The organic nervous system is depressed, by the exposure to fatigue and cold in the first instance, but in the second becomes excited to throw off the evils which imperil it. The pulmonary glands give off a greater amount of vitality or electricity to unite the oxygen of the air with the venous blood, and thus afford fuel to the organic glands for the production of animal heat; hence the burning heat of skin, the strong pulsation of the heart and arteries, and thirst, which supervene, can be accounted for, viz.: the union of the greater quantity of oxygen with the organic glands increases the animal heat; the strong pulsation of the heart is caused by a greater amount of vital fluid being communicated to the nerve-tubules of the cardiac ganglions by the organic glands in the vascular membrane, (which surrounds the ganglions;) the strong pulsation of the arteries is caused by the organic nerves on the internal, middle, and external coats of the arteries being overstimulated, by the excess of oxygen contained in the blood; the thirst is caused by the excess of oxygen in the blood, and water to quench the thirst is demanded, as elsewhere explained, by the spheno-palatine ganglion, to neutralize the excess of oxygen in the blood.

The abstraction of a large quantity of blood removes or diminishes the current of blood for conveying oxygen to the organic glands, and the suspension of animation for a short time reverses the condition of the organic nervous system; instead of being in a state of irritation, it is in a state of depression.

The administration of an anodyne keeps the organic nervous system in a quiescent state; causes contraction of the organic nerves surrounding the arteries; prevents an excess of arterial blood entering them; and consequently, prevents the effusion of lymph, serum, or pus; besides, by arresting the action of the cerebral glands, and consequently the evolition of the volatile agent on whose action through the nerve-tubules of the brain the operations of the mind are made manifest, sleep takes place, and results in the speedy convalescence of the patient.

Hypertrophy of the Heart.
"Digitalis."

The violent action of the heart, induced by hypertrophy of the muscular fibres of the left ventricle of the heart, is generally relieved, and the pulsation of the heart tranquillized, by the administration of the tincture or infusion of digitalis.

Explanation.

The digitalis passes by the lymphatics, lacteals, and thoracic duct, into the venous circulation; it is carried with the blood

10

to the right side of the heart, and from thence to the lungs, where the blood receives its oxygen; it is next carried to tho left side of the heart, with the arterial blood, and is conveyed all over the body by the arteries to the organic glands, to which it is communicated, on the union of the oxygen with them; it is also communicated to the internal, middle, and external organic nerves, on the coats of the arteries, in its transition with the arterial blood through them. The sedative action of the digitalis is communicated to the entire organic nervous system, as well as to the nerve-tubules of the organic ganglions, through their connection with the organic glands in the vascular membrane, which surrounds them: thus the sedative action of the heart and arteries, which follows the exhibition of digitalis, admits of explanation.

Chlorosis.

"Mist. Ferri Comp. Decot.—Aloës Comp.—Nutritious Diet—Porter—Moderate Exercise in the Open Air."

When a girl is observed with a bloodless, waxy expression of countenance, devoid of animation, with languishing eyes and largely dilated pupils; who complains of great debility, and is troubled with palpitation of the heart, as well as a sense of suffocation; whose appetite is depraved, and whose bowels are constipated; in whom menstruation is very scanty, or altogether suppressed, or its place supplied with a leucorrhœal discharge—she is pronounced as laboring under chlorosis, and ordered to take compound iron mixture, with the compound decoction of aloës, nutritious diet, porter, moderate exercise in the open air, and to mix in lively society.

Explanation.

The organic nervous system is prostrated and relaxed; the circulation of the blood is languidly carried on, and the functions of the organic nervous system are impaired. The iron passes into the venous blood through the lacteals and lymphatics; it is conveyed to the right side of the heart, and from thence to the lungs, where the blood receives its oxygen; it is next carried, with the blood, to the left side of the

heart, and from thence, by the arteries, with the blood, all
over the body, to the organic glands, to which it is communicated, on the union of the oxygen with the organic glands;
it is communicated to the nerves on the internal, middle, and
external coats of the arteries, in its transition with the blood
through the arteries; it is communicated to the nerve-tubules
of the ganglions through the organic glands in the vascular
membranes, which are connected with them. The iron imparts strength, vigor, and tonicity to the organic nervous system. The nutritious diet affords material for increasing and
improving the blood. The porter assists the iron in invigor-
ating the organic nervous system. The air supplies good
oxygen for the organic nervous system; whilst exercise increases muscular strength, and lively society renders the animal spirits buoyant.

" Venesection—Organic Nervous System."

Venesection is only proper when there is an excess of oxygen
in the blood, which overstimulates the organic nervous system. viz.: the nerves on the internal, middle, and external coats
of the arteries, the organic glands, and the organic ganglions,
through the connection of the nerve-tubules with the organic
glands in the vascular membrane, which surrounds them.

"When Venesection is Improper."

" In the Stage of Resolution."

As soon as lymph, serum, or pus is formed, the excess of
oxygen in the blood is diminished, and consequently the demand for venesection ceases to be imperative.

" Hepatization of Lung."

In a case of pneumonia, where one of the lungs becomes
hepatized, the withdrawal of blood, and consequently oxygen, would place the patient's life in the greatest jeopardy,
inasmuch as the whole duty of providing oxygen for the blood,
and consequently for the organic glands, would devolve upon
the sound lung, which, under such circumstances, might fail to

accomplish this duty, resulting in the speedy dissolution of the patient, for the want of oxygen to combine with the glands.

"Stage of Collapse."

Venesection is not proper when a person has received a great shock by violence, as a fall from a height, inasmuch as life is nearly extinguished. The pulmonary glands are barely able to evolve as much electricity as will enable enough of oxygen to enter the blood to hold life in the organic glands, which is evident from the fact that the surface quickly gets cold, the pulse almost imperceptible, the countenance pale and ghastly, the respiration feeble. The whole organic nervous system suffers from the want of oxygen and the withdrawal of blood; and consequently oxygen, under such circumstances, would be placing the life of the patient in the greatest peril, inasmuch as it would be taking away the stimulus on which life depended for existence.

"Local Blood-letting."

The local application of leeches, as well as the abstraction of blood by cupping, relieves or prevents inflammation, by withdrawing the oxygen from the organic glands.

"Blood-letting and Stimulants."

The question of blood-letting and the administration of stimulants is one about which there is a very great difference of opinion. In the beginning of the present century, the lancet was invariably and immediately had recourse to on the invasion of any organ by inflammation; the medical men, although untutored in the science of percussion and auscultation, were admirably instructed with respect to the indications afforded by the pulse. As long, therefore, as hardness or firmness of the pulse indicated depletion, they undauntedly persevered in the abstraction of blood, and constantly had their heroism crowned with success, by the convalescence of the patient. It will be recollected that the abstraction of blood, by diminishing the oxygen in the blood, acts as a sedative on the organic glands; and that, by keeping up the sedative action, no mischief could take place, no dilatation of the capillary arteries by the excitation of the organic nerves surrounding them could take place, and consequently, no extra heat, no effusion of lymph, serum, or pus could follow; and consequently, no destruction of the organ attacked could result.

"*Objections to the Administration of Stimulants.*"

If stimulants are given immediately, they most unquestionably add fuel to the fire; they cause more oxygen to pass into the blood; they excite the capillary nerves, causing dilatation of the arteries, the admission of more arterial blood, with an increase of temperature, and the speedy effusion of lymph, serum, or pus, as a matter of course.

"*When Stimulants are Useful.*"

Stimulants are useful in some cases; where, for instance, the organic glands are succeeding, by a new action, in throwing off the excitement, by the effusion of lymph, serum, or pus. Here the stimulants excite the pulmonary organic glands, causing them to give off electricity to combine the oxygen of the air with the venous blood; thus preserving life, whilst the organ affected is undergoing the proper treatment for its restoration to a healthy state.

"*Objections to Indiscriminate Use of Stimulants.*"

The *indiscriminate exhibition of stimulants,* as recommended by the late Mr. Todd, is certainly not called for. Mr. Todd should have qualified his directions, and pointed out the cases where it would be advisable to give them.

Every medical practitioner knows that persons laboring under typhoid fever, typhus mitior or gravior, small-pox, measles, scarlatina, or erysipelas, *will get well without getting any stimulants;* but it does not therefore follow that stimulants are not sometimes required; as, for instance, when the disease under which the patient labors puts on a typhoid character, as indicated by the appearance of the tongue, the action of the heart, and pulse at the wrist. But it requires judgment to decide when stimulants should be GIVEN or when *withheld.*

"Diseases in which Blood-letting should not be employed."

Blood-letting should not be employed in typhoid, *typhus mitior* or *gravior*, small-pox, measles, scarlatina, or erysipelas, inasmuch as these diseases are the result of a specific poison communicated to the organic glands, on the union of the oxygen with them; and further, that the organic glands require a certain number of days to *throw* off the poison; that they very often require assistance, in the shape of stimulants or nutriment, to enable them to do so. Hence depletion at the *commencement* would be jeopardizing the life of the patient at the *termination* of the disease.

Some Agents are followed by different Effects on the Organic Nervous System; so that some Agents can be applied as Antidotes when Death of the Organic Nervous System is threatened with Danger by Poisonous Agents.

" Tobacco—Brandy."

When tobacco, as already explained, finds its way into the arterial blood, it prostrates and nauseates the organic nervous system to the point of death, as indicated by the pale, sunken countenance, the cold surface, the weak pulsation of the heart, the feeble respiration, and extreme exhaustion.

When brandy gets into the arterial blood, it excites and strengthens the organic nervous system, as indicated by the flushed countenance, the strong pulsation of the heart, the hurried respiration, the heat of surface, and buoyancy of spirits.

Brandy, therefore, is the proper medicine to administer when a person is suffering from the poisonous effects of tobacco.

"Strychnine—Tobacco."

When strychnine is introduced into the arterial blood, it causes spasm of the organic nervous system, as indicated by the tetanic spasms of the muscles which supervene.

Tobacco, when introduced into the arterial blood, causes prostration and relaxation of the organic nervous system.

Therefore, tobacco should be administered to a person suffering from poison by strychnine, as recommended by Dr. O'BEIRNE, of Dublin.

" Opium—Green Tea."

Opium, when introduced into the arterial blood, causes contraction of the organic nervous system, and arrests vital action.

Green tea, on passing into the arterial blood, stimulates the organic nervous system, and awakens vital action.

Therefore, green tea should be administered to a person narcotized by opium.

" Hot Water—Cold Water."

When hot water is applied to the surface, as before stated, it irritates the organic nerves, and is followed by heat, pain, redness, swelling, and vesication.

Cold water allays irritation of the organic nervous system, and precludes the occurrence of heat, pain, redness, swelling, or vesication occurring as long as it is applied.

Therefore, cold water is the antidote to hot water.

" Hydrogogues—Opium."

When an overdose of sulphate of soda, jalap, elaterium, or gamboge is administered, and the purgative ingredient passes into the arterial blood, the mesenteric glands become irritated, and secrete the serum of the blood in such quantity as to endanger life, by removing the fluid or blood for conveying the oxygen from the lungs to the organic glands.

Opium causes contraction of the organic nerves in the manner so often described; arrests vital action in the organic nervous system, and, by so doing, interrupts the action of the organic glands, thus arresting the secretion of serum by them.

Opium should therefore be given when hypercatharsis is caused by drastic purgatives.

" Lead—Sulphuric Acid."

Painters, plumbers, and persons in the habit of drinking cider, claret, and beer, are subject to having their organic nervous system poisoned by lead. In the case of painters, the vapor of the lead passes with the oxygen of the air into the blood, and is communicated to the organic glands, on the union of the oxygen with the glands. In the cases of cider, claret, and beer drinkers, the lead gets into the blood through the medium of the lacteals, lymphatics, and thoracic duct, and is given off to the organic glands.

Sulphuric acid, when taken diluted with cold water, passes into the blood, and is also given off to the organic glands, on the union of the oxygen with the glands.

As sulphuric acid combines with lead and forms a sulphate which is inert, it follows, therefore, that sulphuric acid, as recommended by a Scotch miner, is the proper antidote for poison by lead.

" Digitalis—Veratrum Viride—Brandy."

Digitalis, as well as veratrum viride, when administered in an overdose, is followed, on being introduced into the arterial blood and communicated to the organic nervous system, by great prostration and sinking of the vital powers.

Brandy, being followed by different effects, is the proper antidote when the poisonous action of these medicines presents itself.

It is, I presume, not necessary to illustrate the subject further by giving examples, as others will suggest themselves to the reader, and which will admit of explanation in the same manner as that just detailed. (In speaking of antidotes, I purposely avoided alluding to emetics, or the stomach-pump, which, of course, are the chief agents to be employed in cases of poisoning.)

THE MODUS OPERANDI OF MEDICINES

AS THERAPEUTICAL AGENTS

ON THE

ORGANIC NERVOUS SYSTEM.

Brandy — Beef-Tea — Muriate of Soda — Sweet Milk-Whey — Fresh Air.

"*Typhus Fever.*"

A man attacked with typhus fever (in many cases) will be found, on the fourteenth day, lying on his back, sinking down in the bed, with a quick, small pulse; laborious respiration; his tongue covered with sordes, dry, cracked, and contracted; his abdomen tympanitic, having involuntary discharges from the bowels, with either incontinence or retention of urine; besides, laboring under *subsultus tendinum* and derangement of his mental faculties.

In a case of this kind, the patient should be liberally and largely supplied with brandy punch, beef-tea of a rich quality, with a considerable quantity of salt dissolved in the latter; and the room should be kept well ventilated.

Explanation.

The organic nervous system, as indicated by the symptoms enumerated, is prostrated to the lowest degree; the poison has almost destroyed the vital power in the organic nervous system. It is evident, therefore, that the drooping powers of life should be propped up by the exhibition of brandy, which acts as a stimulant on the organic nervous system; by muriate of

soda, which increases the strength of the organic nervous
system; by beef-tea, which affords nutriment to the organic
nervous system, as well as affords nutriment and increases the
supply of blood: thus providing for the renovation of important organs, as well as affording a better medium for conveying the oxygen from the lungs to the organic glands. The
sweet milk-whey increases the quantity of blood. The fresh
air dilutes the poison, as well as affords oxygen of a pure
kind to the blood.

A continuance of the treatment must be guided by circumstances, or until the poison is worn out, when all the functions
of the body will be carried on in the usual way.

"*Punctured Wound of Abdomen—Opium—Ice—Abstinence from Solid Food.*"

When a person has had a dirk-knife plunged into his abdomen, in a very short time after the occurrence he will be found
with cold extremities, weak action of the heart, feeble respiration, and sinking of the vital powers; indicating that the organic nervous system has sustained a great shock.

The treatment should consist in closing the wound, and
giving the patient forty or fifty drops of laudanum, and repeat-
ing the medicine at intervals, so as to keep the patient fully
narcotized for a period of three or four days; abstinence from
food should be insisted upon, and the patient should be allowed ice to quench his thirst.

Explanation.

The opium passes into the venous circulation through the
agency of the lymphatics, lacteals, and thoracic duct; it is carried with the blood to the right side of the heart, and from
thence to the lungs, to be oxygenized; it is next conveyed with
the blood to the left side of the heart, and from thence by the
arteries all over the body to the organic glands, to which it is
communicated, on their union with the oxygen; it is also communicated to the nerves on the internal, middle, and external
coats of the arteries, in its passage through the arteries with
the arterial blood. The opium causes contraction as well as

arrests the vital action of the organic nervous system, thus
keeping the organic nervous system in a perfectly quiescent
state, and causing sleep by arresting the action of the cerebral
glands, and consequently arresting the evolution of the volatile
agent through the nerve-tubules of the brain, on whose action
the operations of the mind depend. The opium arrests the
peristaltic action of the intestinal tube, and contracts it to the
smallest possible diameter, so as to intercept the passage of
the contents of the bowels towards the anus, and thus precludes their extravasation through the wound in the intestine,
(should such an injury have taken place;) in addition, the contracted and steady condition of the bowels keeps the cut surface
of a wound in a quiescent state until inosculation of the nerves
on the cut surfaces of the wound takes place and repairs the injury. The opium prevents inflammation by contraction of the
organic nerves surrounding the capillary arteries, and by arresting vital action in the organic nerves and organic glands.
As lymph, serum, or pus cannot be secreted without the partici-
pation and agency of the organic glands, and without an increase of the calibre of the capillary arteries induced by irritation of the organic nerves, and consequently without an increased quantity of blood and oxygen, it follows as a consequence that the action of the opium in contracting the organic
nerves surrounding the arteries and allaying irritation, by arresting the vital action in the organic nerves and glands, prevents the occurrence of the effusion of lymph, serum, or pus.

The ice quenches thirst without taking in a large draught
of water into the stomach, at a time when it might possibly be
followed by extravasation through a wound in the intestines.
The abstinence from solid food saves the stomach the trouble
of digestion, and the intestines from peristaltic action. A certain time restores the parts to their normal state.

The efficacy of treating peritonitis, caused by perforation of
the intestines, was first pointed out by Drs. GRAVES and STOKES,
in a Clinical Report, published in the 5th volume of the *Dublin
Hospital Reports*, 1830.

Organic Nervous System.

" Practical Remarks."

In the management and treatment of accidents of a formidable character, such as compound dislocation, compound fractures, or laceration of the soft parts by machinery, by which a dreadful shock has been communicated to the organic nervous system, and vitality almost utterly extinguished, the first effort should be, to restore the strength of the organic nervous system, by the administration of stimulants. Amputation should not be performed when a patient is in a state of collapse, after a severe injury, inasmuch as the additional shock would prove destructive to life in the organic nervous system. Reaction should be carefully watched, and as soon as sufficiently established, the necessary operation should be at once performed, to avoid the additional danger to be apprehended from mortification. (Generally, about six hours after the accident amputation should be performed.) As chloroform is followed by depression of the organic nervous system, it would not be prudent to administer it to a person whose organic nervous system is greatly depressed by a severe shock, as it might extinguish its vitality altogether. In a case of this kind, before amputation is performed, the patient should get a free allowance of brandy and water, which would stimulate his organic nervous system, as well as his animal nervous system, and enable him to get through the operation cheerfully.

In cases of profuse flooding after delivery, when the blood has been almost all drained off, not a moment should be lost in having recourse to transfusion, as the patient is dying for the want of blood to convey the oxygen to the organic glands. A person of the same age of the patient, and as nearly as possible of the same constitution, should be selected to supply the blood required. The patient should be made to drink freely of sweet milk, diluted with water, as well as beef-tea, to restore the serum, as well as the fibrin of the blood.

In the treatment of eruptive fevers, amongst which may be classed typhus fever, it should be kept continually in mind that a certain number of days must elapse before the poison is worn out of the organic nervous system; and that the duty of the physician should be to watch the operations of nature, and to give assistance when necessary. As long as a patient goes on without any untoward symptoms, the physician should not be too anxious in his efforts to arrest the disease by active medication, and should recollect, if he keeps the patient alive a given time, the patient will be ultimately able to surmount all difficulties.

In cases where the organic nerves are poisoned by the saliva of a mad dog, or by the virus of an Hunterian chancre, in the former case the part ought to be excised as soon as possible after the part has been poisoned; and the same practice, or the application of caustic potash, (Vienna paste,) should be had recourse to in the latter case. It should be recollected that the poison of hydrophobia may remain dormant for months in the organic nerves of the part bitten, as well as the syphilitic poison continue for a certain period at the site of the chancre, although healed up. It therefore follows that in the case of hydrophobia, if the part implicated is excised at any time before symptoms of hydrophobia set in, that the fearful malady may be averted; and also, if the hard substance which marks the site of an Hunterian chancre is removed, secondary symptoms may be averted.

In the treatment of erysipelas and puerperal fever, depletion should not be had recourse to, if possible; it should be constantly remembered that in these diseases the organic nervous system is suffering from the effects of a poison which will require the action of the organic nervous system for some days to remove, and great caution, therefore, is necessary in the use of the lancet in making long, free incisions, as recommended by Mr. LAWRENCE, in cases of erysipelas, as well as the free venesection in cases of puerperal fever, as recommended by Mr. ARMSTRONG. The loss of blood is accompanied by a corresponding loss of vitality in the organic nervous system, which is thus rendered unable to combat the poison which threatens its very existence.

The treatment of these diseases should not, therefore, be of too heroic a character, but should be rather of an expectant kind, and dictated by the urgency of the symptoms, with a view to their alleviation, as well as to arrest or prevent bad consequences, by proper therapeutic agents.

In the treatment of diseases of a debilitating character, such as phthisis, the object should be to sustain the powers of life in the organic nervous system by the free administration of light, nutritious diet, stimulants in the shape of ale, porter, or beer, exercise in the open air, as well as such other therapeutic agents as are calculated to enrich the quality of the blood and renovate the organic nervous system.

In the treatment of chlorosis, where the structure of the organic nervous system is in a state of inanition, and where the blood is impoverished, the great object to be attained is the rejuvenation of the organic nervous system and the enrichment of the blood. The administration of iron, porter, nutritive animal food, exercise in the open air, agreeable society, conduce to the renovation of the organic nervous system, the restoration of the blood, and the development of the muscular system.

In the treatment of BRIGHT'S disease of the kidneys, when the urine is loaded with albumen, after slightly mercurializing the patient, to dispose of the effects of any chronic inflammation that might be in existence in the organic glands of the kidneys, and next administering iodide of potash to promote a similar action, as well as the absorption of any matter of a foreign character, as also the free use of warm baths, I think it would be advisable to give the patient a table-spoonful of distilled vinegar mixed with water every three hours, as well as to support the constitution with nutritious diet : at the same time, that some preparation of iron should be exhibited. The object in giving the vinegar is to dissolve the albumen contained in excess in the blood; the kidneys could not secrete the albumen, unless provision were made for such a secretion in the blood. The iron invigorates the organic nervous system, which has suffered from the poisonous effects of the urea; and the nutritious regimen restores the blood to a healthy state.

(I may here state, by way of parenthesis, that I am treating two patients on the principles above enunciated, and so far with very favorable results.)

Identity of Erysipelas and Puerperal Fever.

An Immaterial Poison commingled with the Air at certain Periods of the Year, is capable of producing Erysipelas or Puerperal Fever, by poisoning the Organic Nervous System.

It is a well-ascertained fact, that when erysipelas is prevalent, puerperal fever victimizes women recently delivered.

No surgeon will perform an operation, if he can make a choice of the time, when erysipelas is epidemic; he procrastinates, in order to preserve his patient from the complication likely to succeed.

Obstetricians, however, cannot calculate on delay. Pregnant women will bring forth, when the proper time arrives for the uterus to send into the world a new being. It is obvious, therefore, that they cannot impede or guard the internal surface of the uterus from an attack of erysipelas, which, to all intents and purposes, presents the characters of a fresh wound, and consequently is exposed as much to erysipelas as an amputated breast, or any other wound or ulcer on any part of the body. Why, it is reasonable to ask, should the internal abraded surface of the uterus, which HARVEY compared to a stump, after amputation—a *simile* fully coincided in and lucidly demonstrated by Professor SIMPSON—escape with impunity? Does it not communicate with the atmosphere? Is the vagina or os uteri so completely closed as to preclude the admission of air into the cavity of the uterus? Or can it exclude the immaterial poison which is capable of producing erysipelas, and which is commingled with the air, and which is susceptible of being communicated to the organic nerves on the raw surface of the wound, producing diseased action, and giving rise to a group of symptoms called puerperal fever?

What happens when the interior of the uterus is suffering from erysipelas? Why, that the disease extends along the mucous membrane lining the fallopian tubes, and, as there is no obstacle in its path, lays hold of the peritoneum, and further, may embrace the ovaries, through its attachment to them. Again, the erysipelas may pass from the internal to the external surface of the uterus, and thus, by the law of contiguity, involve the serous membrane. It is, therefore, apparent that

160

the peritonitis is merely a continuation of the erysipelatous inflammation, and that the poison extends along the organic nerves distributed to the parts specified, and contaminates them in the manner described.

"The Collections of Pus which are found in different Parts."

How can these pathological phenomena be explained? This question, I presume, can be solved by recollecting what happens to a person who has received a slight wound, on opening the abdomen of a subject who has died of peritonitis. It is well understood that, after a certain interval, the poison is introduced into the system; that great constitutional disturbance is the result; that purulent depositions take place; that the constitution gives way; that the individual in most instances falls a victim to the constitutional irritation excited.*

Explanation.

The venous blood carries the poison from the organic glands to the right side of the heart, and from thence to the lungs, where it receives its oxygen; the blood is next conveyed to the left side of the heart, and is sent by the arteries all over the body; the poison is communicated to the glands, on the union of the oxygen with the glands.

Analogy between the DEBRIS *of the Contents of the Uterus after Parturition, and a Subject recently dead.*

Does not the uterus of a female, after parturition, contain a *débris*, consisting of clots of blood and broken-down decidua? Do not these substances, being extraneous, undergo the process of decomposition, and partake of the character of dead animal matter? Is not the interior of the uterus, when

---

* It should be observed, the wound may be so small as to escape the notice of the individual, as occurred in the case of Mr. DEASE, reported by Mr. COLLES. in the Dublin Hospital Reports, (3d volume,) as well as a similar case given by Mr. TRAVERS, in his work on Constitutional Irritation. It will be further recollected, that only a small vesicle containing serum is found over the wound, and that pus, therefore, is not essentially necessary for the dissemination of the poison through the lymphatics.

161

seized with erysipelas, placed in precisely the same predicament as the hand of the operator in making a *post-mortem*
examination?

When the interior of the uterus is in a state of erysipelatous inflammation, the lymphatics, as well as the veins, come
within its influence; the erysipelas extends to them, the animal poison in which the uterus is bathed finds easy access into
the uterine vessels, and thus into the general circulation, lead-
ing to the purulent deposits, as occurs in the case of the dissecting wound, in consequence of the poison being communicated to the organic glands, on the union of the oxygen with
the former.*

Professor CLARK maintains that puerperal fever is a consequence of endometritis. Dr. ROBERT LEE, that it owes its
origin to uterine phlebitis, with purulent absorption. Now, if
their theory held good, parturient women would be equally
susceptible of taking the disease at all periods of the year.
However, experience and observation contradict, and declare
these *dicta* to be untenable.

The ingenious explanation given by Professor CLARK as to
the connection between the pus found in the ulcerated surface
of the interior of the uterus and in the uterine veins, as well
as the presence of pus in other organs, such as the liver, contrasted with the cases alluded to by Professor SIMPSON, when
no trace of inflammation in the interior of the uterus or its
appendages could be detected, appears very inexplicable.
How is the difference between such eminent men to be recon-
ciled? Each party has undoubtedly recorded the truth. As
there are varieties of erysipelas on the external surface of
the body depending on circumstances it would be superfluous
to enumerate, the same influences, it will be conceded, prevail
where the interior of the uterus is engaged. In the cases
Professor CLARK examined, the erysipelas may have been of a
phlegmonoid character; or poisoning of the organic glands may

* In addition to the erysipelas, animal poison is superadded, derived from the
dead animal matter contained in the uterus. According to the late Mr. COLLES,
there is more danger in dissecting a very *fresh* subject than one in a putrid
state.

11

have taken place there, through the dead animal matter contained in the uterus, which contaminates the organic nerves of the uterus, and afterwards communicates the poison to the venous blood, which is subsequently imparted to the organic glands, on the union of the oxygen with the glands; whereas those that Professor SIMPSON directed attention to may have come under the class of simple erysipelas. Pus is found in the cellular tissue in the former, none in the latter. But is the poison more virulent in one case than in the other? The answer should be, that the forms of erysipelas do not change the constituents of the animal poison, and that the smallest portion of the latter is capable of producing the most disastrous consequences.

In the cases alluded to by Professor CLARK, as well as those mentioned by Professor SIMPSON, the poison was communicated to the organic nervous system. The venous blood carried from the organic glands poisons the venous blood, which, on being arterialized, is conveyed to the organic glands, and conveys the poison to them, in the manner already described.

The next part of the subject appears to be exceedingly intricate: namely, when persons have died of puerperal fever, and whose bodies, on examination, presented nothing remarkable except the black fluid appearance of the blood.* This condition of the blood demonstrates that the vital fluid must have suffered from the effects of a destructive poison. How is this to be accounted for?

The organic glands on the internal surface of the uterus are charged with the poison communicated to them from the

---

* "In three several cases the most careful search was made for morbid alterations of structure, and the lining membrane of the uterus and of the adjacent parts, and nothing could be found to explain the cause of death."—*Dr.* Locock's *Library of Medicine,* vol. i.

"We thus see fatal cases without any proof whatever of omentitis, or any proof of peritonitis, or of metritis, or of uterine phlebitis, or lymphatitis."— *Professor* SIMPSON's *Obstetric Memoirs and Contributions,* vol. ii.

"The morbid changes met with after death are very various. In the most rapidly fatal cases nothing has been met with beyond the mere non-coagulability, thinness and blackness of the blood. The blood in these cases resembles that of persons killed by lightning or hydrocyanic acid."—*Dr.* TYLER SMITH's *Lectures, published in the "Lancet,"* vol. i., 1857.

air. The venous blood is contaminated with the poison derived from the organic glands, which, on being conveyed to the right side of the heart, and afterwards to the lungs, where it receives its oxygen, is next sent to the left side of the heart, and from the latter all over the body, by the arteries, and communicated to the organic glands, on the union of the oxygen with the glands. The poison acts as a direct sedative on the organic nervous system; the pulmonary glands fail to give off electricity or vital fluid. The oxygen ceases to be united with the blood, which becomes of a dark color, and the patient dies, in consequence of the blood not containing oxygen to unite with the glands.

VARIOUS KINDS OF BATHS.

*Modus Operandi of the Warm Bath—Good Effects produced —Prevents Fever and Inflammation—A good Remedy in Strangulated Hernia—Bad Effects of Ice—Remarks on Hernia—Fainting caused by the Warm Bath—Mode of Resuscitation—*MARSHALL HALL'S *Method of Treating Asphyxiated Infants explained—Philosophical Explanation of the Mode in which Infants born apparently dead are restored to Life—Efficacy of Cold Douche in Post-partum Hæmorrhage explained—Mode of Action of Ergot of Rye in arresting Hæmorrhage—How Sprinkling an Infant with Cold Water resuscitates it—How Cold Air or Ice arrests Hæmorrhage after an Operation, when oozing of Blood continues— Wound should not be dressed immediately after an Operation—Good Effects of Cold Douche in Fever and Encephalitis—How a Drink of Cold Water Causes Death—State of a Wound after Exposure for three hours—Mr.* LISTON'S *Remarks—Mr.* MACARTNEY'S *Theory—*SIR ASTLEY COOPER'S *Ideas—Mr.* HUNTER'S *Views—Cold-Water Dressing—Particular Rules with respect to the Mode of Treatment of Wounds after Operations— Effects of the Cold Bath—Explanation of the Modus Operandi of the Cold Bath—Explanation of the Cause of Spasms by Cold—By Tetanus—By Strychnine—By Lead—By Asiatic Cholera—Treatment of Spasms produced by Cold Water— Modus Operandi of Salt-Water Bathing—Phosphorus— Chloride of Sodium—Good Effects of explained—Shower-Baths explained—Cause of Suspended Animation explained —Modus Operandi of Sulphur Baths—Iodine Baths—Nitro-Muriatic Acid Baths explained—Iodine recommended in* BRIGHT'S *Disease.*

WHEN a person is exposed to the vicissitudes of the atmosphere and suffers bodily fatigue, one of the most agreeable remedies he can have recourse to is a warm bath, to recuperate his energy, and relieve the exhaustion and disagreeable sensations he labors under.

It is a question of some importance to know how the warm bath acts in dissipating the symptoms. A person, on being placed in a warm bath circumstanced in the manner just described, will experience a pleasurable or agreeable sensation; on being removed from the bath and placed in bed, he will soon be found in a general perspiration; the warm bath acts as a stimulant on the organic nerves surrounding the capillary arteries; dilatation of the capillary arteries is the result; a greater quantity of blood, with a greater amount of oxygen, thus passes into the capillaries; the organic glands commence to discharge their functions; some of the salts of the blood, with the superfluous oxygen, is united with the hydrogen of the serum of the blood; the latter is decomposed by the electricity evolved at the moment of the union of the oxygen with the organic glands, and forms serum, which passes off by the pores of the skin—the excretory ducts of the organic glands; in this way the sources of irritation to the organic glands are removed, the exciting cause of fever or inflammation is removed by the elimination of the offending matter from the organic glands; thus it will be perceived that a warm bath not only gives comfort in the first instance to the patient, but prevents bad consequences at a more distant period.

In cases of strangulated hernia, a warm bath is a most efficient remedy, if at once had recourse to, as it produces complete relaxation of the muscular fibres, on the same principle as tobacco, chloroform, and venesection; and if a patient is kept in the bath till he faints, reduction of the hernia may be easily accomplished. Whilst speaking on hernia, it may be well to remark, in passing, that the application of ice or ice-water need not be had recourse to if the warm bath fails; and the same remark is true of tobacco, chloroform, and venesection; the cold produces spasm of the capillary arteries, through their connection with the organic nerves, and spasm of the muscular fibres, through the connection of the latter with the nerves, so that the cold only adds to the difficulty, by increasing the intensity of the stricture; therefore, if the warm bath fails, not a moment should be lost in performing the operation for the relief of the stricture. Acting on this rule, I have been very fortunate in operating for inguinal and femoral hernia.

. I wish it to be distinctly understood, that when a man faints in a warm bath, life is temporarily suspended, and all muscular power ceases; and further, that as venesection carried *ad deliquium animi*, is followed by precisely the same effects as fainting in the warm bath, it would be losing time to have recourse to the lancet, on failure of the bath.

Tobacco produces total relaxation of the muscular system, through its deadly influence on the organic nervous system.

Chloroform also produces total relaxation of the muscles, through its action on the organic nervous system; it is evident that all that can be accomplished by venesection, by tobacco or chloroform, is relaxation of the muscular system, and there-fore renders the administration of tobacco, chloroform, and venesection unnecessary.

If a person is kept too long in a warm bath, relaxation and exhaustion of the organic nervous system follow to such an extent, that the person will be deprived of muscular power, the heart will cease to contract, the pulmonary organic glands will cease to give off electricity to unite the oxygen of the air with the venous blood; hence fainting or suspended animation is the result.

What should be the treatment to restore the patient under such circumstances?

Firstly—Immediate removal from the bath into a draught of air.

Secondly—The application of cold brandy to the face, nostrils, lips, body and extremities.

Thirdly—Slapping the hands, feet, cheeks, and other parts of the body, and in addition, a *small* dose of laudanum may be administered. The cold air and brandy cause contraction of the capillary arteries, through the connection of the capillary nerves which surround them, as well as contraction of the muscular fibres of the muscles to which the arteries and nerves are distributed; hence the contraction of the heart can be explained, and the deep sighs or suspiration which follow, in consequence of the heart sending the blood to the pulmonary glands, which stimulates them to give off electricity to unite the oxygen of the air with the venous blood, and by the subsequent union of the oxygen with the organic glands, re-establishing the vital functions.

The slapping causes alternate contraction and dilatation of the capillaries, through the action of the organic nerves surrounding them; the truth of this doctrine must be admitted by every person who has witnessed arteriotomy of the temporal artery, when the blood ceases to spurt. It is well known, twitching the cut extremity with the finger and thumb will cause the blood to flow freely *per saltum,* so that any person can observe the alternate contraction and relaxation of the artery.

The laudanum, in small doses, causes contraction of the capillary arteries through the organic nerves.

The late Dr. MARSHALL HALL stated the marvelous effects of warm and cold baths in restoring animation in an asphyxiated infant. The mode in which the baths act is susceptible of explanation: the cold causes contraction of the capillary arteries through the contraction of the nerves which surround them, and consequently contraction of the muscles to which the arteries and nerves are distributed. The warm bath produces effects diametrically opposite; hence, plunging an infant into cold water and warm water alternately, causes contraction and dilatation of the heart and arteries, and enables the pulmonary organic glands to evolve electricity to unite the oxygen with the venous blood, and thus provide for the resuscitation of life, by the union of the oxygen with the organic glands. The philosophical mode of resuscitating an infant born apparently dead can now be understood, namely: by cold air, warm and cold water baths, slapping the infant smartly, removing obstructions, in the shape of mucus, from the nose and mouth, washing the infant with spirits, as practiced by nurses. The cold air causes contraction of the organic nerves surrounding the capillary arteries; contraction of the muscles to which the arteries are distributed is the result;

contraction of the muscles causes the venous blood to bo forced to the right side of the heart; contraction of the muscular fibres of the heart follows, in consequence of the connection of the organic nerves with the muscular fibres of the heart; the blood is now sent to the lungs; the pulmonary organic glands give evidence that they require oxygen, by the infant opening its mouth and endeavoring to catch air. The organic glands are now stimulated by the air—give off electricity—which unites the oxygen with the venous blood; on the union of the former (the oxygen) with the organic glands, all the functions which characterize life are present. The warm and cold water produces alternate relaxation and contraction of the arteries and muscles, through its action on the organic nerves, and invigorates the action of the capillaries, increasing the force of the circulation. Removing obstructions allows the air to enter the windpipe. The spirits cause contraction of the capillary arteries, through the connection of the organic nerves surrounding them, and thus, contraction of the heart through the nerves which supply its muscular fibres. In connection with this subject, I wish to observe, that when attending as a pupil at the Dublin Lying-in Hospital, I was told that infants supposed to be still-born, and who were placed in locations where their bodies were exposed to a draught of air, were soon afterwards heard crying, and rescued from an untimely grave. Here, the stimulus of the cold causes contraction of the arteries, through the connection of the retina of organic nerves which surround them, and contraction of the muscular fibres through their connection with the arteries, contraction of the muscular fibres of the heart, through the connection of the muscular fibres of the heart with the organic nerves, which are so largely distributed to it; the blood, on being sent to the lungs, stimulates the pulmonary glands, which at once commence their functions, thus restoring life.

In cases of post-partum hæmorrhage, obstetricians are in the habit of dashing cold water on the genitals, with a view of arresting the flooding; and in many instances, this proves a very efficacious remedy. The question as to how the cold water stops the flow of blood is interesting, and one that is now sus-

ceptible of lucid explanation. The cold water, dashed with
force, causes a shock to the organic nerves surrounding the
capillary arteries, causing contraction of the capillary arteries,
and thus prevents the ingress of the blood into them. The capillary arteries, with the organic nerves surrounding them,
supply the muscular fibres of the uterus, so that contraction of
the muscular fibres of the uterus must follow, in consequence
of the muscular fibres of the uterus being supplied with organic nerves; the hæmorrhage, therefore, is promptly arrested
by the contraction of the uterus. It will be recollected that
the uterus, after parturition, presents the character of a recently
inflicted wound, and that the mouths of the capillaries are open.

Ergot of rye arrests hæmorrhage after parturition, precisely
in the same way as the cold douche does; it causes contraction
of the capillary arteries, and through them of the muscular
fibres of the uterus, to which the capillary arteries are distributed, by its irritant operation on the organic nerves surrounding the capillaries.

The ergot of rye acts as an irritant on the organic nerves in
the same manner as lead and strychnine; it is well known that
lead arrests uterine hæmorrhage; and I am satisfied that strychnine would arrest hæmorrhage on the same principle as lead
or ergot does, if properly administered.

The lead causes constriction of the capillary arteries, through
the connection of the organic nerves which surround them.

The ergot causes contraction of the capillary arteries, through
the connection of the organic nerves which surround them.

The strychnine causes contraction of the capillaries, through
the organic nerves which surround them; the lead produces
spasm of the nerves; the ergot produces spasm of the nerves;
the strychnine produces spasm of the nerves; therefore, the
lead, ergot, and strychnine act in the same way as the cold
douche.

Briskly sprinkling cold water on the face of a person who
has fainted, or on the face or chest of an infant in a weakly
state after being born, will cause contraction of the capillary
arteries; contraction of the muscles, through the connection of
the muscles with the arteries; contraction of the muscles,
through the connection of the arteries with the organic nerves;

contraction of the heart, through the shock communicated to the organic nerves surrounding the capillary arteries, which supply the heart; increased action of the pulmonary glands, through the shock communicated to the organic nerves surrounding the branches of the pulmonary arteries.

In cases of amputation of the breast, of the thigh, or the removal of tumors, some surgeons are in the habit of leaving the wound exposed to the cold air for two or three hours, with a view of preventing hæmorrhage; and in some cases where oozing of blood continues after all the arteries have been tied, it is customary to apply either pounded ice or cold water to the bleeding surface.

The *modus operandi* of the cold cannot now fail to be understood: the cold causes contraction of the capillary arteries, through the operation of the organic nerves which surround them; hence no blood can pass through the capillaries, and hence the stoppage of the oozing of blood.

It is easy now to understand how heat produces hæmorrhage, and why it is desirable not to close or dress the wound for some time: the heat acts as a stimulant on the organic nerves surrounding the capillary arteries, causing dilatation of the capillaries, and, consequently, allowing a free entrance of the blood into them, which escapes through their wounded extremities; cold will produce a contracted state of the capillaries, through the shock communicated to the organic nerves surrounding the capillary arteries.

In some cases of fever, as well as in cases of encephalitis, the pain in the head is exceedingly severe; the temporal arteries, as well as the carotids, can be observed pulsating strongly. One of the best methods for relieving a patient so circumstanced, is pouring cold water on the patient's head, which should be continued till relief is obtained, and afterwards had recourse to as often as occasion requires; the cold water causes contraction of the capillary arteries, through the communication of the organic nerves which surround them; hence the quantity of blood sent to the brain is diminished, as well as the heat and pain relieved.

When a person engaged in some violent exercise, and when greatly exhausted, takes a drink of cold water, he may sud-

denly fall dead; the shock produced by the cold on the organic glands of the stomach is at once communicated to the
cardiac ganglion, as well as the pulmonary ganglion; the heart
is rendered powerless as regards its function of contraction, and
the pulmonary glands rendered powerless as to discharging
their function of giving off electricity to unite the oxygen of
the air with the venous blood; life becomes extinct, for the
want of oxygen to unite with the organic glands. It may be
asked, Why a drink of cold water would not produce the same
effect before violent exercise? The answer is obvious—namely
—the organic pulmonary glands have become so exhausted as
to be unable to discharge their functions efficiently any longer;
the additional shock, therefore, of the cold destroys their power
altogether, and places them in the same predicament as if hydrocyanic acid were taken into the stomach; the shock produced by
the cold, under the circumstances, instantly paralyzes the organic nervous system, arresting all the functions of life, and
destroying life itself.

Although it may appear to be a digression, yet I cannot
help alluding to the state of a wound which has been exposed
to the air for three hours. Mr. LISTON, I think, remarks, that
the surface appears glazed and covered with a gray film, and
that, if the surfaces of the wound be now brought in contact,
they will be in a most favorable condition for uniting by the
first intention. I think the glazed condition, as well as gray
film, are attributable to the organic nerves, which, true to
their instincts for the preservation of injured parts, have formed a retina over the entire surface of the wound, with a view
to its protection and renovation; the sides of the wound being
brought in contact, the organic nerves (the *vasa vasorum* of
SIR ASTLEY COOPER) inosculate, and union of the divided parts
is the immediate consequence.

Mr. MACARTNEY states that a wound can unite without inflammation, and the explanation I have given relative to the organic
nerves corroborates the truth of his remark. As a practical
rule, therefore, the wound should not be closed for some time,
nor until the wound should be dry and glazed. Mr. HUNTER
stated that the blood in a wound became organized; but it
certainly acts as a foreign body, and prevents union by the

first intention. No greater improvement in surgery was ever recommended, or based on sounder philosophical principles, than that recommended by Mr. MACARTNEY, of dressing wounds with cold-water dressing; keeping an amputated stump hot, produces inflammation; keeping it cold, arrests inflammation and its consequences; as a practical rule, therefore, it is proper to keep the cut surfaces of a wound exposed to the air for about three hours, until the surface has become dry and glazed, even although there should be no hæmorrhage, or apprehension of hæmorrhage, with a view of healing up the wound in the speediest manner; cold-water dressing is the best application; it is light and cold, prevents the superabundant effusion of lymph, as well as prevents the effusion of serum or pus.

When a person gets into a cold bath, a chilly sensation seizes him; a shock is given to the whole organic nervous system; a feeling of cold is predominant; air is drawn in or inspired, with short and quickly repeated efforts, into the lungs; the surface of the body is made cold by the contraction of the capillary arteries, consequent on the shock by the cold water, to the organic nerves surrounding them, which prohibits the entrance of blood with oxygen into them, and thus shuts off the provision for generating heat by the union of the oxygen with the organic glands. It is necessary, therefore, that some effort should be made to ward off the bad effects induced by the cold, and supply heat; hence, the increased efforts of the pulmonary organic glands to give off more electricity and unite the oxygen of the air with the venous blood, to provide additional fuel for the organic glands, to enable them to resist the shock communicated to them; thus affording one of the strongest proofs of the indivisible connection of the organic nervous glands on the external surface of the body and the pulmonary organic glands; but, notwithstanding the pulmonary glands endeavor to meet the exigency, yet they too are embarrassed, for the cold that has operated on the capillary nerves surrounding the capillary arteries on the external surface of the body causes their contraction; also, by continuity of action, produces a similar state of the capillary organic nerves surrounding the pulmonary capillary arteries; the organic pulmonary glands suffer from the shock, and have their function of giving off electricity suspended; hence the difficulty connected with the respiration, as already stated. After being in the water a short time, the effects of the shock are over. After swimming for some time, the person will be seized with cramps or spasms of the muscles of the lower extremities. To explain this occurrence, it is necessary to remember that the cold, by its shock, has caused constriction of the capillary arteries, through the connection of the organic nerves surrounding and entering into their coats, in the first instance. The continuation of the cold not only causes contraction, but likewise spasm of the capillary nerves surrounding the capillary arteries, thus precluding the entrance of the blood with its oxygen, and thus arresting the provision for heating the surface; the organic nerves surrounding the arteries being distributed to the muscles, and the muscular fibres composing the muscles, it follows as a consequence, that spasm of the organic nerves surrounding the arteries cannot take place without spasm of the muscular fibres taking place, to which these arteries and nerves are distributed. No muscle can be wounded without touching a blood-vessel; therefore, no blood-vessel can be wounded without wounding a nerve; no muscular fibre, without touching an organic nerve; hence, spasm of the organic nerves supplying a muscle is attended with spasm of the muscle also. In corroboration of the truth of the doctrine here propounded, it may be necessary, for the satisfaction of persons doubtful of its correctness, to state, that when strychnine is applied to a blistered surface, when the cuticle has been removed, it will cause spasm of the organic nerves surrounding the capillary arteries; the muscles to which the arteries, with their accompanying nerves, are distributed, are subjected to the action of the strychnine, through their intimate connection with the arteries and nerves. It will be remembered that, when the palm of the hand or sole of the foot is torn by a *rusty* nail, or otherwise suffer laceration from violence, irritation, and eventually spasm of the organic nerves surrounding the capillary arteries, will follow, terminating in the spasm of certain muscles, to which the nerves, circumstanced as stated, are distributed. Again, in painter's colic, the capillary organic nerves surrounding the capillary arteries, distributed to the muscular canal of the intestines; hence, spasm of the intestinal tube is the result. Another instance is afforded by the spasm of the muscles in Asiatic cholera; it is to be recollected that, when a person is attacked with cholera, there is vomiting. with copious discharges of serum from the bowels. In a very short time all the serum of the blood is drained off, the pulmonary organic glands are exhausted, and unable to discharge their function of giving off electricity to unite the oxygen of the air with the venous blood. Again, there is very little blood left to carry the oxygen all over the body; the circulation through the capillaries ceases as a consequence; no blood or oxygen being brought to the organic glands, no heat can be generated; coldness of the surface is the result; the organic glands suffer from irritation; spasm is propagated to the organic nerves surrounding the capillary arteries, to the muscles to which the nerves and arteries are distributed; spasm of the muscles must be the result, and such, in truth, is the fact. It is now proper to ask, What should be the best treatment for a person suffering from cold and spasms produced by immersion in cold water? The answer, after the explanation, must be obvious. In the first place, the person should be thoroughly dried with flannel or coarse cloth—this manipulation implies also friction; the person should be next wrapped up in warm blankets; it should be observed, that if it were practicable to give a warm bath at the onset, it would be the most suitable remedy that could be had recourse to; also a tumbler of hot brandy punch, or, in fact, a hot drink of some kind or other. The philosophy of the treatment scarcely needs explanation. Friction stimulates the organic nerves surrounding the capillary arteries, causing their dilatation, and consequently allowing the entrance of blood, with its oxygen—the provision for heating the surface; the heat causes dilatation of the capillary arteries, through its action on the organic nerves surrounding them, and thus provides for warming the body. The warm brandy punch stimulates the organic nerves of the stomach, stimulates the cardiac ganglion, through the connection of the organic nerves of the stomach with the branches of the par vagum, which latter is connected with the cardiac ganglion, thus increasing the action of the muscular fibres of the heart; the pulmonary ganglion is stimulated also by its connection with the par vagum; thus the pulmonary organic glands are rendered more active in the discharge of their function of giving off electricity to unite the oxygen of the air with the venous blood, and thus make provision for heating the surface of the body. The brandy, as I have elsewhere shown, acts as a stimulant on the organic nervous system all over the body. It causes dilatation of the capillary arteries, through its stimulating effects on the organic nerves surrounding the capillary arteries; thus it will be observed, the brandy increases the muscular power of the heart; increases the power of the pulmonary organic glands; stimulates the organic nerves; makes provision for heating the body, by sending an increased supply of oxygen to the organic glands.

Salt-water bathing is attended with more salutary effects than cold-water bathing. The question arises, Why should this be so? The advantage of the sea-bathing is attributed to different sources: salt water contains a large quantity of chloride of sodium, as well as a considerable amount of phosphorus; the organic nervous system is invigorated by the soda which is brought in contact with it, through the connection of the organic glands with the pores of the skin; the soda thus acts on the pulmonary glands, with which the organic glands of the skin are connected by an indivisible continuity; the pulmonary glands are thus enabled to give off more electricity, and consequently unite a greater quantity of oxygen with the venous blood; hence the rude and healthy blush which characterizes the robust sailor and athletic fisherman. With respect to the phosphorus—it will be recollected that the animal nervous system is spread all over the body in the shape of a retina; the ultimate filaments of the animal nerves enter into the formation of the retina. No one can doubt this explanation, who pricks any part of the body with a pin, as he will feel pain; the phosphorus is brought in contact with the animal nervous system by immersion in the salt water; the nerves are stimulated by the phosphorus. It is well to remark, that when a man studies hard, phosphorus will be found in the urine, so that it is evident the brain or animal nervous system suffers, so far as the phosphorus is concerned; the phosphorus of the salt water supplies its place. Thus, it will be observed, sea-bathing invigorates the body and rejuvenates the mind.

When a person takes a shower-bath, there is a shock communicated to the whole organic nervous system; the capillary arteries all over the body are contracted through the shock communicated to the organic nerves surrounding them; the muscles are contracted in consequence of their connection with the arteries and nerves; the frequent inspiration shows that the pulmonary nerves are affected, and evidently shows that they experience the shock; if the shower-bath be continued too long, the patient will fall in the bath; contraction of the capillaries will be followed by spasm of the capillaries and organic glands; the pulmonary organic glands will fail to give off electricity, and animation will be suspended, for the want of oxygen to unite with the venous blood; hence it is that a man will fall under the violence of a shower-bath, as if pierced with a bullet through the heart. I should state, that spasm of the muscular fibres of the heart will also be produced by the continuation of the shock caused by the shower-bath; so that the action of the heart being suspended, death follows, as an inevitable sequence, no blood or oxygen being sent to the glands; the immaterial agent known as life cannot continue the occupant of the organic nervous system; the vital spark becomes extinguished for the want of oxygen, just as a galvanic battery would cease to give off electricity for the want of acid, or as a fire would cease to burn when totally excluded from the air. When a person leaves the bath, and is thoroughly rubbed with a coarse cloth, he will soon experience a glow of heat, and his skin will present a red surface all over: the friction stimulates the nerves; the nerves quickly recover from the shock, and act with increased vigor; the chastisement they have received from the shower-bath whets their energy for circulating the blood through the capillaries; the circulation is therefore rendered more vigorous, animal heat is increased, strength of the muscles is insured, and the individual rendered buoyant to the highest degree; relaxation of the muscles, with loss of power, cease to render an individual a burden to himself. Such are the good effects of shower-baths for persons adjudged to require them.

The sulphur bath kills the animalculæ by the sulphur which passes, incorporated with the water, through the pores of the skin.

The iodine bath acts on the organic glands of the skin precisely as if administered by the mouth, and exercises its specific action on the testis, the mammary, parotid, or other glands.

I wish to remark, that when iodine is administered by means of a bath, or by external application in the way of ointment, or internal administration by the stomach, or internal administration in the shape of vapor by the lungs, as soon as its physiological agency on the organic nervous system is established, the urine, on the addition of starch, will assume a violet color; it therefore follows, as a consequence, that the *corpora malpighiana*, or organic renal glands, are charged with iodine; the iodine, therefore, must act on the organic nerves of the kidney. As BRIGHT's disease, or granulated kidney, so called, in consequence of the appearance presented by making a section of it, is characterized by the large quantity of albumen in the urine, as well as the absence of urea, would it not be advisable to give *large doses of iodide of potassium*, to act as an antidote, and eventually exterminate the latent poison that has induced "*Morbus Brightii?*"

The nitro-muriatic acid bath acts on the organic nervous glands of the skin, and through these glands, on the organic nerves surrounding the capillary arteries of the liver, enabling this organ to discharge its functions more efficiently, in the elimination of the bile from the blood. The nitro-muriatic acid gives tone and vigor to the nerves, as well as the organic glands; hence the beneficial results which follow the use of these baths in chronic enlargement of the liver, accompanied with jaundice.

MODUS PROPAGANDI OF THE HUMAN SPECIES.

Modus Propagandi of the Human Species—Sexual Excitement—Animal Nervous System Internuncio between Mind and Genital Organs—Organic Nervous System—Mode of Distribution in Genital Organs—Connection between Animal and Organic Nerves—Genital Organs—State of Genitals previous to and after Coition—Propagation of Life—Old Men should not Marry—Bad Effects from—Masturbation—Spermatorrhœa—Treatment—Young Married Men—Over-Sexual Indulgence—Seminal Emissions—Convulsions—Impotence—Treatment—Impotence from Moral Causes—Treatment—Prostate Gland demonstrated to be the Muscle instituted for the Expulsion of the Semen.

WHEN a passion of an amatory character seizes a buoyant young man, ushered in by the presence of a female to whom he is attached, peculiar excitement of the animal and organic nervous systems is the immediate result, presently announced, by the orgasm of the genital organs; the picture impinged on the retina is seen by the mind, through the medium of the nerve-tubules of the optic nerves; the impression passes through the nerve-tubules of the brain, through the nerve-tubules of the spinal cord; to the nerve-tubules of the spermatic nerves, which arise from the spermatic ganglions, to the cremaster muscles and the testicles, through the connection of the cremaster muscles and testicles with the organic nerves surrounding the spermatic arteries, as well as through the pudic nerves, branches of the sacral plexuses, to the prostate gland, levatores ani, WILSON's muscles, erectores penis, compressores venæ dorsalis penis, corpora cavernosa penis, corpus spongiosum penis, and glans penis.

From what has now been stated, it is evident that free communication takes place between the mind and genital organs, and that the operations or wishes of the mind are conveyed to all the organs to which the nerves are distributed, as the brain, or mind, located in the brain, extends to them through the nerve-tubules of the nerves.

It is now proper to point out the operation of the organic nervous system. It is to be remembered that the animal and organic nervous systems act in harmony; the brain, or the mind, located in the brain, communicates with the superior central organic ganglion, through the crura of the brain attached to it; the ganglion communicates through the brain, and par vagum, with the solar plexus, or, more correctly, with the spermatic ganglion, and spermatic plexus, derived from the spermatic ganglion, which accompanies the spermatic arteries, in their destination, and final distribution in the testicles; the pudic artery is surrounded by a plexus or retina of nerves derived from the hypogastric plexus; the spermatic nerves freely communicate and inosculate with the organic nerves surrounding the spermatic arteries, thus establishing a free communication between the two sets of nerves; the pudic nerve takes the same course as the pudic artery, and is distributed to the prostate gland, levatores ani, and WILSON's muscles, erectores penis, compressores venæ dorsalis penis, (HOUSTON's muscles,) corpora cavernosa penis, corpus spongiosum penis, and glans penis, and inosculates with the organic nerves surrounding the pudic arteries, thus establishing a free communication between the animal and organic nerves of those parts, or between organic life and animal life.

This short description of the arrangement of the animal and organic nerves in the genital organs, now given, will enable the student to proceed to a further examination of what takes place during sexual excitement.

On the brain or mind communicating with the organic nerves distributed to the cremaster muscles and testicles, the cremaster muscles are thrown into action, and the testicles are elevated or approximated towards the external abdominal rings, in order to shorten the distance between the vasa deferentia and prostate gland; the organic nerves surrounding the spermatic arteries become expanded; dilatation of the arteries is the result; through the intimate connection of the former with the latter, a greater quantity of blood, with a greater supply of oxygen, is the sequence; the organic spermatic glands formed at the termination of the capillary arteiers commence to secrete the semen, which is carried away by the seminal ducts. The process by which the semen is formed is well worthy of consideration. The increased quantity of blood supplies the material for the production of the semen; the increased quantity of oxygen is for the purpose of increasing the vital power of the organic glands, in the production of the seminal element—the union of the oxygen with the glands is accompanied by increase of temperature and the evolution of electricity; when, therefore, the gland has elaborated the peculiar element of the semen, in accordance with its function, under the guidance of the spermatic ganglion, the electricity decomposes some of the serum of the blood whilst circulating through the gland, the hydrogen of which unites with the seminal element and oxygen, with which it is in combination, and thus forms the seminal fluid, which is conveyed by the seminal ducts towards the vas deferens. It will be remembered that water, which is composed of oxygen and hydrogen, forms the great bulk of the seminal fluid; at the same moment that communication is had with the testicles, other correspondence is had with the prostate gland, levatores ani, WILSON's muscles, erectores penis, compressores venæ dorsalis penis, corpora cavernosa penis, corpus spongiosum penis, and glans penis, through the medium of the pudic nerves, which correspond and inosculate with the organic nerves accompanying the branches of the pudic arteries; the organic nerves surrounding the arteries distributed to the prostate gland become strong, firm, and contracted; the arteries to which they are connected become similarly circumstanced, as well as the muscular fibres and cellular tissue to which the arteries are distributed; the prostate becomes firmly contracted; the *canals* of the common seminal ducts which enter the prostate at its *base*, and open near its *apex*, are thus rendered impervious; the

organic nerves distributed to the levatores ani, WILSON's muscles, the erectores penis, and compressores venæ dorsalis penis,
are similarly circumstanced: hence, the same condition of the
arteries and muscles follows—namely, firm contraction; the
organic nerves distributed to the arteries of the corpora cavernosa, corpus spongiosum, and glans penis, become at first dilated,
followed by a corresponding dilatation of the arteries—thus
admitting a large quantity of blood to enter their trunks and
capillaries; the penis now becomes firm and turgid, whilst, at
the same time that the movement of the blood in the arteries
is taking place, the compressores venæ dorsalis penis are discharging an important duty. It will be observed, unless some
barrier were placed to the return of the blood by the veins,
that the arteries would be soon unloaded, and that flaccidity of
the penis would follow—particularly, on any attempt at intromission into a narrow vagina: hence it is that the thin tendon
of the compressores venæ dorsalis penis, forming a flat band
over the dorsal vein, prevents the return of the blood, and
keeps the penis erect; indeed, the penis may be truly said, when
in a state of erection, to be surrounded by a firm band at its
root, formed by these muscles; therefore, as long as the muscles continue in this condition, the penis must continue firm,
and in a state of erection, the organ being previously injected
with blood.

Intromission of the penis into the vagina is immediately followed by a motion on the part of the male, attended with friction or titillation; the testicles, during this time, are secreting
the semen, which is carried by the vasa deferentia towards the
common seminal ducts, which permeate the prostate gland;
but, as the canals are closed by the contracted condition of
the gland, the semen regurgitates into reservoirs called the
vesiculæ seminales; as soon as these reservoirs can contain
no more, and commence to press on the prostate gland, the
" *vis-à-tergo* " is followed by a shock that pervades the whole
frame, and the semen is ejected with force, and *per saltum;*
the friction of the penis in the vagina, (just like tickling the
soles of the feet, will, after some time, cause alternate contraction and relaxation of the organic nerves surrounding
the arteries, followed by a similar state of the arteries them-

182

selves, and the muscles to which the arteries are distributed,)
in due time, is followed by spasm of the organic nerves; the
arteries and muscles to which the nerves are distributed induce
relaxation and contraction of the muscular fibres of the prostate gland, levatores ani,* WILSON's muscles, erectores penis,
compressores venæ dorsalis penis; thus, it is at the proper moment the gland or muscle dilates, when the semen rushes into the
common seminal ducts, the gland or muscle then suddenly contracts, the semen is ejected with proportionate force, the muscle again expands and contracts, with similar results, until all
the semen is discharged. Cotemporaneously with the movement of the prostate, a similar movement of the levatores ani,
WILSON's muscles, the compressores venæ dorsalis penis, and
erectores penis takes place: alternate relaxation and contraction of these muscles assist the prostate in the expulsion of the
semen; the alternate relaxation and contraction of the levatores ani propel the semen from the vesiculæ seminales into the
common seminal ducts, and empty them; WILSON's muscles, by
alternate relaxation and contraction, eject the semen forward;
the erectores penis promote the same movement; the compressores venæ dorsalis penis render assistance by a similar action;
again, by this action of the muscles, pressure is taken off the
veins, the blood is allowed to return to the general circulation,
and the penis, worn out by exertion, drops, pendulous and flac-
cid, into a quiescent state.

   It is a remarkable, and a very important matter, to watch
what occurs when the semen is about being discharged. It is
true that, towards the latter end of the process of coition, the
respirations become shorter and shorter, until respiration is
momentarily suspended, just as the discharge of semen is about
taking place. The *suspension* of respiration is *isochronous*
with the shock *communicated* to the *whole frame* on the *emission* of the semen; the electricity or vital fluid given off by the
pulmonary organic glands, to unite the oxygen of the air with
the venous blood, has its operation for a moment interrupted,

* These muscles may be all looked upon as one pair of muscles, viz., the *levatores ani*. The muscles called after their discoverers are merely divisions of
the levatores ani.

it being necessary that such should take place in order to propagate life to another individual; therefore, the electric or *vital*
fluid, instead of *being given off* by the pulmonary glands for the
purpose of continuing life in the usual way, has its course
changed, and directed to the semen, just being discharged.*
Thus it is that man imparts a portion of his own life to his offspring; hence it is that derangement of the organic nervous
system is communicated from one individual to the other—as,
for instance, epilepsy; hence it is that the offspring of drunk-
ards are often afflicted with epilepsy, and other diseases of a
nervous character.

The semen being now placed in the vagina of the female,
charged with the vital agent in the manner just described,
having an affinity for the ovule located in the ovum, attracts it
from the ovary into the uterus, where union takes place between the semen and the ovule, and lays the foundation for
the formation and organization of the future individual.† The

---

* The Creator demonstrated, by what appears to be a material act on His
part, the mode in which life should be communicated by man to the semen:
"And breathed into his face the breath of life"—(Chap. ii., v. 7, *Genesis;*) thus
showing the process by which man should continue to propagate his species.

† It would be criminal to cause abortion at this period. It is the general belief that, during a fruitful coition, the ovule embraces the semen immediately.
Most women are conscious of conception the moment it takes place. It would
be, therefore, sacrificing life to produce abortion under the circumstances. The
Saviour remained nine months in the womb of the VIRGIN MARY, as is found
from *St. Luke*, 1st Chap., 35th verse:

"And the Angel answering, said to her, the Holy Ghost shall come upon
thee, and the powers of the Most High shall overshadow thee; and therefore,
the Holy which shall be born of thee shall be called the Son of God."

Persons skeptical about the truth of the text just quoted must admit that
there is no more difficulty in understanding its meaning, than in understanding
the manner in which a loadstone is capable of rendering a bar of steel magnetic,
or impregnating it with an invisible and imponderable agent of extraordinary
power by simple contact.

On the Origin of Species by Means of Organic Affinity. By H. FREKE,
A.B.M., B.T.C.D.M., R.I.A., Fellow of Kings and Queens College of Physicians, &c., &c.

For a review of the above work, I am indebted to the *American Medical
Times.* The author says in the Preface, "Nothing is advanced in this publica-
tion that is not perfectly in harmony with the Mosaic Record of the Creation."
It is evident, when the learned author speaks of "chemical affinities," "organic

semen, when charged with the vital agent, is guided by the same or similar laws as those which govern a magnet, that will attract a piece of steel, or particles of steel, in its immediate vicinity, for a given period, and no longer; thus showing that its influence is capable of extending a certain distance, and no farther. In like manner, the miniature of the internal and external organization of the male is thrown on the semen; and in like manner, the vital agent is capable of forming and continuing to increase the organization of the body for a certain time, and no longer. To demonstrate that a portion of the vital agent, or Life itself, is imparted to the semen during the process of coition, or sexual communication, I will endeavor to prove this fact by a familiar illustration. When a broken-

---

affinities," "organizing atoms," "the organizing residual product," "specific stimulus," "origin of species by means of organic affinities," "the embryo of organic creation," "one parent of all since existing organic creation, the other part being, as I conceive, a mineral or inorganic world," that he does not believe the first chapter of Genesis, verse 24: "*And God said, Let the earth bring forth the living creature in its kind, cattle and creeping things and beasts of the earth according to their kind, and it was done.*" Assuming for a moment that all the ingenious theories of the author are correct, a difficulty presents itself, which appears to overthrow them—a matter which will be confessed by every person who has dissected the human body, who has appreciated its organization, and studied the uses of the various organs—namely, how the first animal of each species was formed. If the embryo of organic creation possessed such consummate wisdom as is displayed in the organization of the creeping creature, and the still greater wisdom displayed in the organization of man, it is evident it possessed greater wisdom, greater mechanical ingenuity, greater knowledge of chemistry, greater knowledge of physics, greater knowledge of acoustics, greater knowledge of optics, greater knowledge of harmony, greater knowledge of physiology, than any person now possesses. It is certain that man, with all the means at his disposal in this world, could not make a model representing every organ in the human body perfectly; and it is equally true, with all the chemical affinities at his disposal, he could not impart life to any object that did not possess vitality before. Therefore, materialists are not able to explain the phenomena of animal organization and vitality, inasmuch as they are not able to make a single living animal. DARWIN maintains that all organic creation proceeds from a "PRIMORDIAL CELL." The same arguments present themselves to his theories. It is impossible to conceive how a "CELL" should be possessed of such intelligence as to make a man!!!

" Credat Judæus Apella, Non Ego."

Man, in the words of ARISTOTLE, is a $\mu\iota\chi\rho\sigma\varsigma$ $\chi\sigma\sigma\mu\sigma\varsigma$, or *little world*, beyond the comprehension of chemists or materialists.

down old man, rendered youthful in appearance by an artistic
hair-dresser, a scientific dentist, and fashionable tailor, contracts marriage with a dashing young widow, his appearance,
as well as the train of symptoms that soon present themselves,
leave no doubt that he is regularly "used up." His pale
countenance, sunken eyes, feeble or tremulous gait, pain in the
loins, accompanied by palpitation of the heart, hurried respiration, loss of appetite, as well as the hypochondriasis he labors under, indicate the wreck of his organic nervous system.
The cause of all those troubles can be easily explained. Every
time the old fellow, to gratify his vanity more than his passion,
has connection with his wife, he gives off a portion of his life;
continued destruction of the vital agent is soon followed, not
only by the symptoms above described, but convulsions or
sudden death; the organic pulmonary glands become so exhausted, that, at length, they are unable to give off enough of
electricity or vital fluid to unite the oxygen with the venous
blood—death is therefore caused by the want of oxygen to
combine with the organic nervous glands and ganglions. Old
men should, therefore, "look sharp," and ponder well before
they get themselves entrammeled in the troublesome bonds of
wedlock with a widow.

The treatment calculated to restore to health a person suffering from the disturbance of the organic nervous system,
produced by the causes specified, commands attention. The
patient should be sent on a visit to some distant place, where
he would be removed from the embraces of his wife; as soon
as his dyspeptic symptoms are removed, he should be liberally
supplied with animal food, and get a fair share of malt liquors;
the administration of some preparation of iron and nux vomica
would be advisable, as well as the alternate use of warm and
cold shower-baths, carriage exercise in the open air, pleasant
society, &c.

The same group of symptoms which characterize the troubles of a feeble old man, when he gets married to a lascivious
widow,* are found in a young man who has practiced, and

* The widow of a robust young man never thinks of the disparity between
the age of her first husband and the old man to whom she gets married a second
time. Her motto is, " Tros Tyriusque mihi nullo discrimine agetur."

continues to practice, masturbation. The expression of timidity, the languishing or suspicious eyes, the pallid features, the palpitating heart, the offensive breath, the flatulent stomach, the pain in the small of the back, the frequent micturition, the many and numerous ills, the patient will inform you, that harass his mind and body, will at once point out the genital organs as being the *"fons et origo mali."* On inquiry, the patient will tell you he has either practiced masturbation, or that he is in the habit of doing so.

There is no more troublesome disease to treat than spermatorrhœa; the patient the subject of it is invariably a hypochondriac, and liable to fall a victim to quacks and impostors. The patient is to be truly pitied, and his case should be thoroughly understood, with a view to its alleviation or cure. Instead of frightening the wits out of the unfortunate sufferer, as is very often done by unprincipled persons, who love money more than they do the patient, and make the miseries of the latter subservient to the accumulation of the former, the patient should be assured, by attending to the directions given him, that he would be restored to health and vigor within a given time—varying from three to nine months. When the organic nervous system is weakened to such a great degree, the patient, on being fully enjoined to desist *in toto* from the practice of self-abuse, should be ordered nutritious animal food, malt liquors, iron combined with nux vomica, conium, shower-baths, lively society, and sea-bathing; he should be made to get up early in the morning, and to attend to his usual business.

I am well aware the use of stimulants and animal food is countermanded by able surgeons; but, when the condition of the patient is taken into consideration, their utility will be at once recognized. In a patient reduced to the low state I have . described, the blood is impoverished, and not present in sufficient quantity; the tissue or structure of the organic nervous system is deteriorated and attenuated; therefore, it is necessary to restore the one and repair the other. In consequence of the condition of the blood and organic nervous system, the prostate gland is in a feeble and relaxed condition; the slightest irritation of the penis will be followed by the discharge of mucus from the follicles connected with the prostate gland, or, sometimes, seminal fluid, if deposited in the vesiculæ seminales. In the commencement of masturbation, the organic nerves will hold out for a considerable time before they contract and relax; but, when the practice is continually persisted in, they yield almost immediately, and relaxation and contraction are the immediate results, with the discharge of fluid. The animal diet, with the porter, iron, and nux vomica, increase the quantity of the blood, as well as improve its quality; and, further, provide for the renovation and invigoration of the organic nervous system—rendering the latter firm, strong, and buoyant. The substance of the organic nervous system suffers from wear and tear, the same as the other organs of the body: the iron furnishes the blood with the particular element required for the restoration or renovation of the substance of the organic nervous tissue; whilst the nux vomica gives strength and energy to the nerves; as is exemplified when the prostate gland firmly contracts, closes the seminal ducts, and prevents the passage of the semen through them into the urethra: the conium acts as a sedative on the organic spermatic glands of the testicles, and prevents or arrests the secretion of semen: the warm baths, followed by cold shower-baths, regenerate and strengthen the organic nervous tissue; change of air and employment invigorate the mind. I have not alluded to the cauterization of the seminal ducts, as practiced and recommended by LALLEMAND, as I deem such treatment totally useless—(1 speak from experience on the matter)—without attending to the constitutional treatment; merely treating the effect, without removing the cause, is not satisfactory. The chief advantage to be derived from the caustic is the moral effect it produces: it causes a sufficient amount of pain to deter the patient from manipulating the virile organ, and thus inducing excitement or irritation of the genital organs. LALLEMAND advises the administration of half a pint of cold water, as an enema, at bedtime, with good results: the cold water causes the contraction of the muscular fibres of the prostate gland; thus closing the common seminal ducts, and thus precluding the passage of the semen from the vesiculæ seminales into the urethra. It is, therefore, to be remarked, that the application of caustic alone cannot be relied on, without attending to the constitution. The reason why it is desirable to prevent the secretion of the semen is, to prevent the irritation of the genitals which follows, and to give time to the prostate to recover its strength and energy.

Mr. HUNTER states that, if a man continues to take opium for a certain period, he becomes impotent. The same remark is confirmed by SIR ASTLEY COOPER. The opium, therefore, interferes with and prevents the secretion of the semen by the spermatic glands in the testes. Conium is given on the same principle as opium acts.

Young married men, who have lived virtuously previous to their marriage, will, after some months, complain of a train of symptoms, which will leave no doubt of their having indulged too freely in their marital rights. After advising the party to indulge less in sexual intercourse, and to avoid stimulants in the shape of wine, ale, or porter, the patient takes his departure, and after the lapse of some weeks or months returns, and says he has adhered to the rules laid down, but that he is no better, and is still " miserable and wretched." On inquiry, the patient will tell you, after he passes water, that something like starch follows, and that a considerable quantity of starchy matter flows from the penis during the process of defecation. How is this matter to be explained? The man adheres to his instructions, has no connection with his wife; but his mind contemplates the act; the result is, the semen is secreted and lodged in the vesiculæ seminales; the muscular fibres of the prostate not being sufficiently strong, but partially relaxed, the pressure of the fæces on the vesiculæ seminales forces the semen through the common seminal ducts, no opposition being given by the muscular fibres, as just explained. The continued discharge of the semen accounts for the symptoms. The patient should get conium, iron, nux vomica, cold-water enemata, be sent away from his wife, and be allowed nutritious diet, with malt liquors. It is true that, in persons who practice masturbation, as well as young married men, the penis, at the commencement, will bear a great deal of friction before alternate contraction and relaxation of the organic nerves are induced, followed by alternate contraction and relaxation of the arteries, as also the muscles to which the arteries are distributed; but it is equally true that, after a certain time, debility will set in, and that the powers of resistance will, as already explained, give way, as evidenced by the convulsions which in many cases supervene: hence, the alternate relaxation and contraction of the organic nerves are not confined to a part of the organic nervous system, but attack the whole organic nervous system, and shake life to its very foundation. Convulsions, in truth, may be deemed a struggle between life and death, under whatever circumstances they occur; therefore, they must be always looked on with apprehension. Convulsions are attributable to, or caused by, a want of oxygen to combine with the organic nervous ganglions and glands, inducing spasm or alternate contraction and relaxation of the organic nerves surrounding the capillary arteries, and a similar condition of the arteries, and of the muscles to which the arteries are distributed. Therefore, when convulsions attack a man addicted to self-abuse, it must be recollected that vitality has been previously impaired; that the whole organic nervous system has been subjected to repeated shocks; and that, consequently, the pulmonary organic glands are partially paralyzed, and incapable of discharging their functions—namely, giving off electricity or vital fluid to combine the oxygen of the air with the venous blood, which is to be subsequently conveyed by the arteries to the organic ganglions and glands, to keep in existence the spark of life. That this is the true explanation of the exciting cause of convulsions, must strike every person with its truthfulness, who has ever witnessed the convulsions of an animal bled to death—the convulsions which set in before death takes place, when all the blood is drained off, show the struggle life makes for oxygen before its departure from its abode in the organic nervous system.

In connection with the evils resulting from masturbation, it is well to state, that sometimes complete loss of nervous power in the organic nerves takes place; the man has the will to have sexual intercourse, but is unable to obtain an erection. In this instance, repeated abuse has caused paralysis of the organic nerves—a matter sometimes proved by the paraplegia which accompanies it. The paralysis of the lower extremities is caused by the diseased condition of the nerves extending along the course of the pudic to the internal iliac, common iliac, and aorta. The organic nerves surrounding the arteries having their vital functions destroyed, accounts for the paralysis of the lower extremities which follows. The impotence induced by this cause should be treated by the administration of animal food, such as beef, mutton, oysters, together with ale or porter, iron combined with strychnine, small doses of opium, cold shower-baths, sea-bathing, horse exercise, the introduction of bougies, mixing in lively society, and abstaining from tobacco in every form. I have not mentioned electricity, as it is not *applicable* to diseases of the organic nervous system; it kills the vital agent in the organic nervous tissue *whence* it supplies the place of the immaterial agent in the animal nervous system. In corroboration of this doctrine—(see experiments detailed by Brown-Séquard in his work)—the form of impotence here described must not be confounded with that form alluded to by Sir Astley Cooper, when a young man fails to have an erection through too great a desire to accomplish his purpose.

It is now established, I presume, that certain moral causes, as well as certain physical agents, will cause extreme debility and relaxation of the organic nervous system, incapacitating it from discharging its functions: as, for instance, a sudden fright will stop the functions of the pulmonary organic glands, followed by fainting; the pulmonary glands being unable to give off electricity, or the vital fluid, to combine the oxygen with the venous blood to be subsequently conveyed by the arteries to the organic ganglions and glands; suspended animation is the result. Tobacco, when taken into the stomach, will be followed by similar effects. Again, certain moral causes and physical agents will induce a vigorous state of the organic nervous tissue. Buoyancy, induced by good news of great import to the individual's future happiness in life, will be accompanied by a vigorous condition of the organic nervous system; the eyes will sparkle, the countenance will be animated, the muscles will be invigorated, the person will "leap for joy." Again, a physical agent, in the shape of a tumbler of French brandy punch, will induce an excited state of the organic nervous system.

To revert to the patient who suffers from impotency from overanxiety, follow Sir Astley Cooper's advice. Enjoin him on no account to attempt to have sexual intercourse for some months with his wife; order bread-pills to be taken regularly three times a day; the impression made on his mind with respect to the matter will counteract the other previously made; and just recommend, by way of elevating his spirits, in consequence of your prohibition, the imbibition of a tumbler of punch before going to bed, and the case will give no further trouble, as the organic nerves will discharge their duties immediately.

It sometimes happens that ripe bachelors and men of strictly moral habits, who lead lives of celibacy, but who indulge in luxurious or idle habits, as well as gratify their appetites with a considerable amount of animal food, and, besides, drink ale, porter, wine, or punch, to make them happy and jovial before going to bed, have the mortification as well as discomfort, on awaking from a dream, to find themselves surrounded by damp linen, in consequence of profuse seminal emissions. Here, it is to be remembered, that Sir Astley Cooper says, a man in his health will have a seminal emission every ninth day, and that the mode of living above described promotes such an occurrence. Although persons can command their passions while awake, they cannot control the operations of the mind when in a dream; the animal propensities will conquer the moral under such circumstances.

" Si expellas naturam furca, recurrit atque recurrit."

When consulted in a case of this kind, as will occasionally happen, the man should be put on moderate diet, all stimulants should be prohibited, bodily exercise insisted on, as well as early rising; the bowels should be kept free. A short persistence in this course of treatment will soon set all matters right.

As some persons may think I attribute too much importance to the prostate gland, it is meet and right I should give as much satisfaction as possible to persons skeptical on the correctness of the doctrine with respect to it I have put forward on the present occasion. It is to be recollected, the prostate gland is not found developed in young children, or in boys, until the age of puberty. It is also to be remembered, the prostate becomes degenerated in old age; its functions and development are therefore coeval with the epoch a man is to propagate, and continue to propagate his species. Such being the case, it follows that the prostate must play an important part in the propagation of the species. The importance of its office may be succinctly explained. The testicles represent a distilling apparatus for the secretion of the semen; the vasa deferentia represent, in *shape* and *construction*, two worms connected with the stills, for carrying the liquor or semen to the receivers, the vesiculæ seminales; the prostate gland, when contracted, as it is during the process of coition, furnishes a stop-cock—the common seminal ducts are closed by it; the vesiculæ seminales furnish receivers for the semen to be collected in, until required to be discharged through another pipe, (the urethra,) to its final resting-place in the vagina. It is evident that, if the semen requires to be sent to a distant part, that a pump is required for the purpose; and such a mechanical contrivance is furnished by the prostate gland. The gland or muscle dilates and contracts in the same manner that the heart contracts and dilates. After a given period, the semen is accumulated behind the prostate gland, but it cannot pass through, in consequence of the contraction of the gland or muscle; the gland or muscle now dilates, the semen rushes into the ducts, the gland now forcibly contracts, the semen is expelled by force and *per saltum;* an interval takes place; again dilatation takes place, and again contraction, with another discharge of semen; and so on until the entire semen is discharged. It is evident, therefore, that Mr. THOMPSON's account of the prostate gland being muscular is correct; and let me state, that next to the heart, it is the most *important* muscle in the body; that, in truth, it is the propagating muscle, or a muscle without which the species could not be propagated. The harmony of action between the levatores ani, WILSON's muscles, the erectores penis, the compressores venæ dorsalis penis, and the prostate, is wonderful. The adaptation of means to accomplish ends so amazingly constructed and arranged, thus presented by the examination of the important subject I have attempted to elucidate, indicates the wisdom of the Omnipotent Creator, and must strike the student with wonder at every step of his investigations.

SYPHILITIC POISONING

OF

ORGANIC NERVOUS SYSTEM.

Description of the Mode in which a Man contaminated by Syphilitic Poison communicates the Syphilitic Poison to a Healthy Woman—The Mode in which a Healthy Woman communicates the Disease to the Fœtus in Utero—The Mode in which the Infant communicates the Disease to a Healthy Nurse—The Mode in which the Nurse communicates the Disease to the Fœtus in Utero—The Mode in which a Nurse communicates the Disease to a Healthy Child.

It is an exceedingly difficult matter to eradicate true syphilitic virus when once firmly ingrafted on, and fully communicated to, the organic nervous system. The chancre or primary ulcer may be followed by ulcerated tonsils, by copper-colored scales over the head, trunk, and extremities, accompanied by iritis, at *a remote period*—perhaps between five or seven years; hydrosarcocele of the testicle, with thickening of the periosteum, and iritis, may again present themselves, accompanied with necrosis of the nasal bones. In the event of the individual contracting marriage in the mean time, he will communicate the disease to his wife, who, in turn, will communicate the disease to the fœtus *in utero*. The latter, if born alive, will be found covered with copper-colored blotches on the nates; and after some time, will be observed to be covered with scales, which will soon run into ulcers; the infant at this time presenting an emaciated, shriveled appearance, closely resembling a *monkey*, troubled with snuffles, and giving vent to a peculiar shrill cry. An infant so circumstanced, on being applied to the nipple of a healthy nurse, will communicate the disease to her, and in due time she will be observed to be covered with a copper-colored eruption; and in the event of her being pregnant, the disease will be communicated to the fœtus *in utero*, which will present the characteristics of *syphilis* in due course, on being *born*.

I will now explain briefly how the state of things described is brought about. The semen secreted by the testicles contains the syphilitic poison; the semen, on being brought in contact with the organic glands, on the *cervix uteri*, communicates the poison to them; the venous blood conveys the poison to the right side of the heart; on the blood being oxygenized, and conveyed to the left side of the heart, it is distributed by the arteries to the organic glands; the poison is communicated to the glands, on the union of the oxygen with the glands. In the event of there being a fœtus *in utero*, the poison is communicated to the organic glands in the placental lobules; on the blood being oxygenized, it is conveyed by the umbilical vein to the fœtus, when it passes into the arteries, and is given off to the organic glands of the fœtus, on the union of the oxygen with the glands; thus the organic nervous system of the fœtus is poisoned. Again, when an infant thus suffering from syphilis is applied to the nipple of a healthy nurse, the poison is communicated from the lips of the child, which generally will be found in an ulcerated state, to the organic glands of the nipple; the syphilitic poison is in this manner communicated to the nurse, who in due time affords proof of inoculation by the poison. It is to be further remarked, that the nurse, when thus contaminated by the syphilitic poison, will communicate the poison to a healthy child: the mammary glands secrete the milk; the milk must be impregnated with the poison, inasmuch as it is derived from a poisoned source; the child *sucks* the nipple; the milk passes down into the stomach; the milk is absorbed by the lacteals and lymphatics, and passes by the thoracic duct into the venous circulation; on the blood arriving at the right side of the heart, it is sent by the pulmonary artery to the lungs, where it is oxygenized; it is next sent to the left side of the heart by the pulmonary veins, from whence it is distributed by the arteries all over the body. The poison contained in the blood is communicated to the organic glands of the fœtus, on the union of the oxygen with the former; in due time the child will present the characteristics, as well as symptoms, that it is suffering from syphilis.

With respect to the syphilitic poison, which is capable of generating syphilis in the various phases, as well as circumstances enumerated, it is the one called after Mr. HUNTER, and is the one which the eminent Mr. CARMICHAEL used to declare in his lectures "could not be eradicated without the administration of mercury."

Having already described the manner in which mercury, when applied externally or administered internally, finds its way into the circulation, and is communicated to the organic glands, on the union of the oxygen with the glands, it is merely necessary to state that the mercury, by its more powerful agency, neutralizes or destroys the power of the syphilitic poison contained in the glands, or rather the syphilitic poison by which the glands are contaminated, and that, on the action of the mercury being kept up, the organic glands cease to give evidence of the presence of the syphilitic poison.

Experience, however, demonstrates to a certainty, that it is impossible to predict when a sufficient quantity of mercury has been administered to kill the disease, inasmuch as the disease known as secondary or tertian syphilis presents itself when least expected. Indeed, the evils resulting from the poison of syphilis are not confined to the person who is originally contaminated with it, but are propagated to his offspring, giving them an unhealthy aspect, as well as ingrafting on their constitutions all those disagreeable features and contingencies which are grouped under the term *scrofula*.

Small-Pox.

The Mode in which Small-Pox is communicated by a Woman laboring under Small-Pox to the Fœtus in Utero.

When a pregnant woman who has not been vaccinated is attacked with small-pox, in the event of her being in the last months of pregnancy, she will give birth to a child covered with pustules, showing that the disease has been propagated to it whilst *in utero*. The venous blood of the mother conveys the poison to the right side of the heart, from whence it is conveyed by the pulmonary artery to the lungs, where it is oxygenized. The blood is next conveyed by the pulmonary veins to the left side of the heart, from whence it is transmitted by the arteries all over the body, and communicated to the organic glands in the placental lobules. The organic glands, thus poisoned, communicate the poison to the blood, which, on being arterialized, is carried by the *umbilical vein* into the arterial circulation of the fœtus. The poison is communicated to the organic glands of the fœtus, on the union of the oxygen with the glands. The result is, the formation of papillæ, vesicles, and pustules on the fœtus, precisely in the same manner as if the poison, by a slight incision, were applied to the organic nerves on the skin of the arm. Thus, what is considered, and what most assuredly is, an extraordinary phenomenon, is susceptible of explanation.

Nævi Materni.

The Mode in which a Mother communicates Impressions of certain Objects presented to her View to the Fœtus in Utero.

In accounting for certain marks impressed on the fœtus *in utero*, in consequence of objects presented under peculiar circumstances to the mind through the nerve-tubules of the retina, the nerve-tubules of the brain, the nerve-tubules of the spinal cord, the nerve-tubules of the spermatic ganglions, which communicate with the nerve-tubules of the spinal cord, the nerve-tubules of the organic nerves which surround the ovarian arteries, the nerve-tubules of the arteries distributed to the placenta, which communicate with the nerve-tubules of the nerves which surround the hypogastric arteries of the fœtus, the impression of the object presented to the mind through the nerve-tubules of the retina is directly communicated to the fœtus *in utero*, through a continued chain of communication, viz., through the nerve-tubules of the animal and organic nervous systems, until it is eventually impressed on the fœtus *in utero*.

13*

Animal life, or what is called the mind, is coextensive with
the nerve-tubules; the white matter contained in the nerve-
tubules is of the same character as the white matter contained
in the nerve-tubules of the brain; the white matter is the seat
of the mind, precisely as the gelatinous matter contained in
the tubules of the organic ganglions, glands, and nerves, is the
seat of life in the organic nervous system. Life and animal
life act in unity and unanimity; whatever disturbs one equally
disturbs the other; whatever object is visible to one becomes
also visible to the other.

The object presented to the mind through the nerve-tubules
may be said to communicate with the fœtus *in utero* through
the nerve-tubules of the nerves, which connect it with the
nerve-tubules of the retina. If a man looks through a long
cylinder, he can observe a man's face at the extremity of the
cylinder, whilst the man at the extremity can also observe the
other, who is looking at him through the cylinder; just in the
same way the mind can look through the cylinders of nerve-
tubules at the fœtus *in utero*, whilst the fœtus *in utero* can
communicate through the same cylinders with the object which
is presented to it, and take its impression. The image of an
object presented to the mother's eye can be daguerreotyped on
the fœtus.

Explanation of the Mode in which Parts of the Body, such as the Nose, the Ear, or the Fingers, when removed by Accident or Design, are reunited.

"Inosculation of Organic Nerves."

Authenticated histories of numerous cases are on record
where either the *nose*, the *ear*, or a *finger*, has been removed,
and yet has been restored by judicious treatment, which consisted in applying the part removed carefully and exactly to
the place it belonged to, and retaining it for a certain number
of days in that position, or until union has taken place.

It must be exceedingly interesting not only to the surgeon,
but the physiologist, to be able to explain how this extraordinary phenomenon is brought about.

I will endeavor to present the reader with what I am fully convinced is the true solution of the mystery.

It is a well-known fact that the heart of a reptile, on being removed from the body, will leap several feet from the ground, and continue to do so for some time.

Every person who has observed a butcher skinning an ox or a sheep, must have remarked the twitching of the muscles— what the butchers call the "*fish.*"

Again, it is true that an eel may have the head cut off, the skin peeled off, and be divided into two or three parts, which, on being placed on the gridiron, will spring off the latter into the fire. The phenomena just described are attributable to *vitality* still existing in the organic nerves of the organ or parts of the body alluded to.

When the nose is *cut off* by a sword, *vitality* continues in the *organic nerves* of the part removed for some time; precisely in the same way that vitality continues in the heart of the reptile, or in the muscular fibres of the ox, or in the divisions of the eel.*

Therefore, when the cut surfaces of the wounded part are regularly approximated and retained in contact, the organic nerves will inosculate; the mouths of the capillary arteries must inosculate as a matter of course, in consequence of the inosculation of the retinæ of organic nerves surrounding the capillary arteries; circulation of blood through the arteries follows, accompanied by animal heat, and in due time all the functions of the part will be restored.

In connection with this matter, it will be observed that Dr. WARREN, of Boston, remarks that, when the Rhino-plastic operation is performed successfully, the individual will refer any irritation of the part of the nose that has been borrowed to the place from whence it received it, and consequently will refer the trouble to the forehead; and if it hap-

---

* In case the ox, the reptile, or the eel has been killed by hydrocyanic acid or an electric shock, no leaping of the heart of the reptile takes place; no movement of the muscles of the ox will be perceived; no writhing of the eel is perceptible, for the obvious reason that vitality is expelled from its abode in the organic nervous system. It will be remembered, the abstraction of blood does not destroy vitality for some time.

pens to be the Taliacotian operation, he will refer the itching of the tip of the nose to the part of the arm the tip of the nose originally belonged to. The important part the nerves play is rendered manifest, inasmuch as, if the union took place through the effusion of lymph, or through what Sir A. Cooper calls the *vasa vasorum*, irritation would not be referred to the location from whence the new nose was removed, but be confined to the organ itself.

As further evidence that the organic nerves are the instruments through which union is established, it may be stated that practical surgeons make it a rule to use fine ligatures, and to tie the ligatures sufficiently firm to divide the middle and internal coats of the artery. The utility of thus securing the artery is susceptible of other explanations besides those generally given by authors, viz.: that on the division of the middle and internal coats of the artery by the ligature, retraction of these coats instantly follows; the organic nerves of the external coat of the artery, being torn from their attachment to the elastic coat of the artery, are exposed, and, on being brought in contact, inosculate; thus completely obliterating the artery, and rendering the part above the ligature in appearance similar to a cord.

The wounds made in cases of hare-lip, or cancer of the lip, are united by the inosculation of the organic nerves of the raw surface. I will reiterate, with the late Mr. Macartney, that inflammation is not necessary for the union of a wound by the first intention; that inosculation of the nerves is all that is required. It is to be further observed, that the largest description of wounds will unite, if properly adjusted, viz., by simply approximating the edges of the wound, and barely holding them in contact, avoiding all supernumerary appendages in the shape of dressing. In proof of the correctness of this doctrine, I subjoin a note from the *learned* and observant Dr. John Watson, of this city:

"New York, *November* 18th, 1861.

"Dear Doctor—You are correct in reference to union by the first intention and without suppuration, in the case upon which I operated in Brooklyn; but as to the number of days, you have made too rapid a recovery. I have on several occasions succeeded after amputation in closing the wound by the first intention, but this is the only one in which the success was perfect on the thigh: the others were amputations either of the arm, forearm, or leg.   Yours truly,

"JNO. WATSON.
"Dr. Jno. O'Reilly."

Union of wounds by the first intention is always aimed at, in cases of·amputation, by the Irish as well as British surgeons.

In conclusion, I cannot help observing, that if any additional proofs were required to demonstrate the necessity of understanding the laws which regulate the organic nervous system, it would be afforded by the practice of the *French* surgeons, who still continue *to retard* the speedy convalescence of patients after amputations, by *filling* the *stumps* with *charpie;* thus subjecting the patients to a very *tedious* process of cure, and one, too, very *often* attended with *fatal* consequences.

The French surgeons are brilliant and dashing operators, and the great mortality which follows their operations, when contrasted with those of American surgeons, must be attributed to the after-treatment, being conducted on totally different principles by the latter.   I do not wish it to be understood that I am anxious to elevate the character of American surgeons at the expense of their French brethren—the duties of humanity alone suggest the comparison, and the hope is ardently entertained that the latter will appreciate the truly scientific treatment of the former.

Remarks on Dr. Brown-Séquard's *Work on the Physiology and Pathology of the Central Nervous System.*

Dr. C. E. Brown-Séquard states that section of the cervical sympathetic is attended with the following phenomena:

1. *Dilatation of the blood-vessels.*
2. *Afflux of blood.*
3. *Increase of vital properties.*

Which he attributes to paralysis of the *blood-vessels*, causing *more blood* to pass through the vessels *in a given time*, producing an *increase* of the *vital* properties of the contracted and narrow tissues. He says that himself, Drs. WALLER, DONDERS, and others, have proved by experiment that whatever may be the cause of *increasing* the *circulation* of blood in the blood-vessels of the head in *a given* time, produces almost all, if not all, the phenomena following section of the cervical sympathetic.

Galvanization of the cervical sympathetic causes—

1. *Contraction of blood-vessels.*
2. *Diminution of blood.*
3. *Decrease of vital properties.*

Dr. BROWN-SÉQUARD says the result of the experiments by section of the nerves, as well as by galvanization, shows " the *untenability* of a *vitalistic* theory, according to which the normal actions of the sympathetic nerve would be *increased* after it has been divided and diminished, when it is excited by galvanization; and according to which, also, nutrition and animal heat would be dependent upon the sympathetic nerve, which would produce an increase of these two functions after it had been divided, *(although it ought then to cease to act;)* and a *diminution of these functions* when it is galvanized, *(although* it then ought to *act more than normally.")*

Dr. BROWN-SÉQUARD's experiments, although they resulted diametrically opposite to what he expected, and were followed by effects contrary to what in his mind should be anticipated, yet prove conclusively the doctrine I have propounded. The section of the nerve *irritates, not paralyzes*, the sympathetic nerve; hence the phenomena described by Dr. BROWN-SÉQUARD were just what should be expected from irritation of the nerve.

With respect to the application of galvanism to the nerve *decreasing* instead of *exciting* the powers of the nerve, the result was such as might be anticipated, on recollecting the galvanic current killed the vital agent in the nerve precisely as electricity or lightning destroys life by the shock communicated to the organic nervous system. The cervical nerve is an organic or vital nerve, and differs in this respect from an animal nerve, that may have a galvanic current sent through it with impunity.

Dr. BROWN-SÉQUARD asks, " What is the origin of the cervical sympathetic nerve ?" Dr. BROWN-SÉQUARD answers, that he conceives, with Dr. WALLER and Prof. BUDGE, that the nerve-fibres of the cervical sympathetic that go to the *iris* *originate* from the *spinal cord.* But this description is met by Dr. QUAIN's description of the organic nervous system in the acephalous fœtus, as well as by the comparative anatomy of the Invertebrata, which have no cerebro-spinal nervous system.

It is to be remarked, that what Dr. BROWN-SÉQUARD calls PARALYSIS is in truth IRRITATION. What he calls IRRITATION is in truth PARALYSIS. *Section,* or *wounding* the nerve, causes *irritation.* *Galvanism,* or electricity, *destroys* the vital agent in the nerve; hence the *deadly* shock which follows.

With respect to Dr. BROWN-SÉQUARD's observation that the blood-vessels of the head are supplied with nerves " *chiefly* from the *spinal cord,*" " by the roots of the *last* cervical and *first* and *second* dorsal nerves," I apprehend he will find very *few* anatomists to agree with him on this point. *All* anatomists, with the exception of himself, agree that the blood-vessels of the head are supplied with nerves, which form a retina round them, from the cervical organic ganglions. Dr. GRAINGER has shown that the ganglions are connected by two roots to the anterior and posterior pillars of the spinal cord—the anterior and posterior roots of the spinal nerves. The same arrangement of the animal and organic nervous systems is observed here as is followed all over the body. The animal nervous system is spread out at the termination of the nerves into a net-work, which *inosculates* with the organic nervous system, at the termination of the capillary arteries.

Remarks on Professor SIMPSON's *Statement that there are no Nerve-Fibres in the Umbilical Cord.*

In the foreign correspondence of the *American Medical Times* for January 19th, 1861, will be found a letter from DAVID P. SMITH, M.D., in which the following paragraph occurs: " Professor SIMPSON regards the fact that there is no nerve-fibre in the umbilical cord, as proof positive that there can be no influence exerted by the imagination of the mother upon the fœtus in utero."

Professor SIMPSON's name must always command respect and

attention, but a man cannot deny the evidence of his own senses. I distinctly recollect the particulars of four cases where impressions made on the minds of the mothers were conveyed to the fœtus in utero. In one case, a woman in the last month of pregnancy had presented to her the dead body of her husband, who had been kicked to death by his horse. In some days afterwards she was delivered of a son, who, when I saw him at the age of three years, presented all the characteristics of an idiot. In another case, a woman witnessed a frightful accident befall her husband during the last days of her pregnancy; in due time she was delivered, and the child, a daughter, when I saw her, being over three years old, had the characteristics of idiocy, as well as a convulsive movement of the muscles resembling paralysis agitans.

In a third case, where a woman in the last month of pregnancy had been robbed of the hard earnings of herself and husband, which she had deposited in a chest, sustained a tremendous shock, rendering her almost powerless to do anything for some time, was shortly afterwards delivered of a son, which is now perfectly idiotic.

In a fourth case, a woman sustained a great fright, in consequence of her husband having received injuries by machinery; she was soon afterwards delivered of a child, which she brought to me at the age of three months, in consequence " of a beating of the heart." I examined the little patient, and, with the exception of the increased action of the heart, there appeared to be nothing whatever wrong.

P. S.—I have corrected the last sheet for the press on this evening, the 21st November, 1861; and as I entered on the investigation of the placenta on the evening of the 21st November, 1858, it is exactly three years since I commenced to study the contents of this volume.

www.ingramcontent.com/pod-product-compliance
Lightning Source LLC
Chambersburg PA
CBHW021708210326
41599CB00013B/1568